Progress in Molecular and Subcellular Biology

Series Editors: W.E.G. Müller (Managing Editor), Ph. Jeanteur, I. Kostović, Y. Kuchino, A. Macieira-Coelho, R. E. Rhoads

32

Springer

Berlin
Heidelberg
New York
Hong Kong
London
Milan
Paris
Tokyo

Progress in Molecular and Subcellular Biology

Volumes Published in the Series

Volume 13	**Molecular and Cellular Enzymology** Ph. Jeanteur (Ed.)
Volume 14	**Biological Response Modifiers: Interferons, Double-Stranded RNA and 2′, 5′-Oligoadenylates** W.E.G. Müller and H.C. Schröder (Eds.)
Volume 15	**Invertebrate Immunology** B. Rinkevich and W.E.G. Müller (Eds.)
Volume 16	**Apoptosis** Y. Kuchino and W.E.G. Müller (Eds.)
Volume 17	**Signaling Mechanisms in Protozoa and Invertebrates** G. Csaba and W.E.G. Müller (Eds.)
Volume 18	**Cytoplasmic Fate of Messenger RNA** Ph. Jeanteur (Ed.)
Volume 19	**Molecular Evolution: Evidence for Monophyly of Metazoa** W.E.G. Müller (Ed.)
Volume 20	**Inhibitors of Cell Growth** A. Macieira-Coelho (Ed.)
Volume 21	**Molecular Evolution: Towards the Origin of Metazoa** W.E.G. Müller (Ed.)
Volume 22	**Cytoskeleton and Small G Proteins** Ph. Jeanteur (Ed.)
Volume 23	**Inorganic Polyphosphates: Biochemistry, Biology, Biotechnology** H.C. Schröder and W.E.G. Müller (Eds.)
Volume 24	**Cell Immortalization** A. Macieira-Coelho (Ed.)
Volume 25	**Signaling Through the Cell Matrix** A. Macieira-Coelho (Ed.)
Volume 26	**Signaling Pathways for Translation: Insulin and Nutrients** R.E. Rhoads (Ed.)
Volume 27	**Signaling Pathways for Translation: Stress, Calcium, and Rapamycin** R.E. Rhoads (Ed.)
Volume 28	**Small Stress Proteins** A.-P. Arrigo and W.E.G. Müller (Eds.)
Volume 29	**Protein Degradation in Health and Disease** M. Reboud-Ravaux (Ed.)
Volume 30	**Biology of Aging** A. Macieira-Coelho
Volume 31	**Regulation of Alternative Splicing** Ph. Jeanteur (Ed.)
Volume 32	**Guidance Cues in the Developing Brain** I. Kostović (Ed.)

Ivica Kostović (Ed.)

Guidance Cues in the Developing Brain

With 17 Figures

Springer

Professor Dr. Ivica Kostović
Croatian Institute for Brain Research
School of Medicine
University of Zagreb
Šalata 12, 10000 Zagreb
Croatia

ISSN 0079-6484
ISBN 3-540-00117-4 Springer-Verlag Berlin Heidelberg New York

Library of Congress Cataloging-in-Publication Data.
Guidance cues in the developing brain / Ivica Kostovic (ed.).
p. cm. – (Progress in molecular and subcellular biology ; 32)
Includes bibliographical references and index.
ISBN 3540001174 (hardcover : alk. paper)
1. Developmental neurophysiology. 2. Axons. 3. Nerves–Growth. 4. Telencephalon. 5. Molecular neurobiology. I. Kostovic, Ivica, 1943– II. Series.

Springer-Verlag Berlin Heidelberg New York
a member of BertelsmannSpringer Science+Business Media GmbH
http.//www.springer.de

Printed in Germany

Production: PRO Edit GmbH, 69126 Heidelberg, Germany

Typesetting: SNP Best-set Typesetter Ltd., Hong Kong
Cover design: design & production, 69126 Heidelberg, Germany
39/3130/göh – 5 4 3 2 1 0 – Printed on acid-free paper

Preface

Many complex molecular interactions are involved in the development of mammalian brain. Molecules serving as guidance cues for migratory cells, growing axons and recognition of postsynaptic targets have recently become a major topic for research because they are directly involved in the formation of neuronal circuits, which enable the subsequent functional development through interactions with the environment. In addition, most molecules acting as guidance cues are also involved in plasticity, damage repair and regeneration in the adult brain.

This volume reviews current knowledge about major classes of molecules involved in the guidance of growing axons; tau proteins involved in the establishment of axonal polarity, morphology, and outgrowth; gangliosides and lectins involved in neuronal migration, neurite outgrowth, and contact recognition; and myelin molecules that inhibit nerve regeneration.

In the introductory chapter, Judaš et al. review current research on the spatio-temporal distribution and function of the four major classes of axonal guidance cues (netrins, semaphorins, slits, and ephrins) and their receptors in the developing mammalian telencephalon. These guidance cues have an overlapping distribution in the specific architectonic zones of the cerebral wall during the embryonic and early postnatal period, as well as complementary and/or overlapping functions in the development of all major classes of telencephalic axon pathways: subcortical (thalamic and extrathalamic) afferent systems, corticofugal (projection) systems, and corticocortical (commissural and ipsilateral) fibre systems. However, most axonal guidance cues have other developmental roles as well, including their possible involvement in synaptic plasticity in the adult brain.

Šimić et al. outline the current knowledge about the neuronal cytoskeleton, microtubule-associated proteins, and growth cone translocation. The authors focus on tau gene organization and regulation, the tau protein structure, and discuss how the subcellular localization of tau phosphorylation modulates its proposed roles in the establishment of axonal polarity, morphology, and outgrowth. The available evidence suggests that changes in the phosphorylation state of tau serve as a mechanism whereby transduction pathways of different extracellular and intracellular signals lead to distinct changes of the growth cone cytoskeleton.

Rösner describes the developmental expression and possible roles of gangliosides in brain development, as revealed by means of biochemical, immunohistochemical, enzymatic and genetic approaches. Increasing evidence suggests specific roles of 9-*O*-acetyl-GD3 in neuronal migration, of GM1 in calcium ion homeostasis and axonal outgrowth, of GM2 in pyramidal neuron dendritogenesis, and of GT1b in neuron/oligodendroglia interaction promoting myelin stability and controlling nerve regeneration.

There is increasing evidence that lectins are widely distributed in the nervous tissue of mammals. Zanetta reviews the involvement of mannose-binding lectins in the development of the cerebrum, focusing on their role in cell adhesion mechanisms and in the stimulation of myelinating cell proliferation. Mannose-binding lectins seem to play crucial roles in contact guidance of neuronal migration, interneuronal recognition, myelination and tightening of the ependymal cell barrier.

Mendez-Otero and Cavalcante describe their studies on functional role of gangliosides in neuronal motility. They identified and characterized a specific glycolipid, the 9-*O*-acetyl-GD3, localized to the membrane of neurons and glial cells that is expressed in regions of cell migration and neurite outgrowth in the developing and adult nervous system. The ganglioside 9-*O*-acetyl GD3 could provide a new cell-signaling mechanism in glial-guided neuronal migration and has been implicated in both neuronal migration and neurite outgrowth. The potential roles for this ganglioside in identical mechanisms for neuronal migration and neurite outgrowth suggest an emerging framework in which glycolipids are involved in cell motility in general.

Finally, Schnaar examines the limits of nervous system plasticity by focusing on myelin molecules that inhibit nerve regeneration, their nerve cell surface targets, and the signal transduction pathways that they trigger. Three inhibitory molecules (Nogo, myelin-associated glycoprotein, and chondroitin sulfate proteoglycan) bind to complementary target molecules on the nerve cell surface, thereby initiating a signal transduction cascade that results in growth cone collapse and a halt in axon outgrowth. Thus, myelin inhibitors profoundly restrict recovery from CNS injury. An understanding of these molecular interactions and signal transduction pathways may provide new therapeutic approaches to enhance nerve regeneration after injury.

Zagreb, Spring 2003

Ivica Kostović

Contents

Complex Patterns and Simple Architects: Molecular Guidance Cues for Developing Axonal Pathways in the Telencephalon
M. Judaš, N.J. Milošević, M.-R. Rašin, M. Heffer-Lauc, I. Kostović

Complex Patterns and Simple Architects: Molecular Guidance Cues for Developing Axonal Pathways in the Telencephalon

M. Judaš, N.J. Milošević, M.-R. Rašin, M. Heffer-Lauc, I. Kostović

1 Introduction

The aim of this review is to discuss the current research on the spatio-temporal distribution and function of the four major classes of axonal guidance cues (netrins, semaphorins, slits, and ephrins) and their receptors in the developing mammalian telencephalon. In the first part, we briefly describe guidance molecules and their receptors. In the second part, we review their overlapping distribution in the specific architectonic zones of the cerebral wall during the embryonic and early postnatal period. In the third part, we describe complementary and/or overlapping functions of these molecules in the development of all major classes of telencephalic axon pathways: subcortical (thalamic and extrathalamic) afferent systems, corticofugal (projection) systems, and cortico-cortical (commissural and ipsilateral) fiber systems. To conclude, we discuss several general themes which emerge from the current research, and point out that most axonal guidance cues have other developmental roles as well, including possible involvement in synaptic plasticity in the adult brain.

2 The Four Major Classes of Axonal Guidance Cues Are Netrins, Semaphorins, Slits, and Ephrins

Growth cones of developing axon pathways successfully navigate through the complex environment of the fetal telencephalon to their intermediate and final targets, by responding to a variety of substrate-bound or diffusible molecular signals (Tessier-Lavigne and Goodman 1996; Mueller 1999). These molecular guidance cues on cell surfaces and

Croatian Institute for Brain Research, School of Medicine, University of Zagreb, Šalata 12, 10000 Zagreb, Croatia

Progress in Molecular and Subcellular Biology, Vol. 32
Ivica Kostović (Ed.)

in the extracellular matrix (ECM) are expressed in stereotypical patterns and either positively or negatively regulate the direction and speed of axon growth and the degree of axon fasciculation (Letourneau et al. 1994; Goodman 1996; Mueller 1999). The molecules anchored to the cell membrane or localized to the ECM surrounding the cell that secretes them serve as local guidance cues and influence the behavior of growth cones via contact-mediated mechanisms. Soluble molecules secreted by the intermediate or final targets of growing axons serve as long-range guidance cues, and growth cones advance up or down a concentration gradient of a particular guidance cue (Goodman 1996; Mueller 1999).

Axonal guidance cues represent a complex set of signaling molecules: cell surface receptors and adhesion molecules, nondiffusible ECM molecules and short-range ECM-bound guidance cues; and diffusible long-range attractants or repellents (Tessier-Lavigne and Goodman 1996; Chisholm and Tessier-Lavigne 1999). Many of the molecular guidance cues and their receptors are evolutionarily conserved, displaying similarities in the structure and functional roles in organisms as different as a nematode (*C. elegans*) or a fruitfly (*Drosophila*) and mammals, including humans (Chisholm and Tessier-Lavigne 1999). The four major families of recently identified axonal guidance molecules include the netrins, semaphorins, slits, and ephrins (Culotti and Kolodkin 1996; Flanagan and Vanderhaeghen 1998; Chisholm and Tessier-Lavigne 1999; Mueller 1999; Raper 2000; Mellitzer et al. 2000).

2.1 Netrins Usually Function as Chemoattractants and Bind to Deleted in Colorectal Cancer and Neogenin

The netrins, a small family of secreted laminin-related proteins, were the first diffusible chemoattractant molecules identified in the vertebrate central nervous system (Kennedy et al. 1994; Serafini et al. 1994). Receptors for netrins are Deleted in Colorectal Cancer (DCC) and the neogenin (Mueller 1999; Chisholm and Tessier-Lavigne 1999). The netrins are expressed along the ventral midline of the CNS in both invertebrates and vertebrates, and they attract commissural axons to the midline (Colamarino and Tessier-Lavigne 1995; Culotti and Kolodkin 1996). Mutations in mouse netrins result in defects in commissure formation in both the spinal cord and the forebrain (Serafini et al. 1996). In the spinal cord, Netrin-1-responsive commissural axons express cell adhesion molecule TAG-1 as they approach the midline, and cell adhesion molecule L1 after they cross it (Dodd et al. 1988). However, in the forebrain, commissural axons retain the expression of both the TAG-1 and the L1 after they cross the midline (Fujimori et al. 2000). Netrins are also chemoattractants for cerebellofugal axons in the hindbrain (Shirasaki et al. 1995). However, netrins can function as chemorepellents for certain classes of axons – for example, axons of the trochlear nerve are repelled away from the midline by netrin (Varela-Echevarria et al. 1997).

2.2 Semaphorins Act as Chemorepellents for Most and Chemoattractants for Some Axons and Bind to Neuropilins, Plexins, and L1

The semaphorins act as chemorepellents for most axons and as chemoattractants for some growing axons (Mark et al. 1997; Raper 2000). The semaphorins are a large family of transmembrane and secreted molecules, currently classified into eight groups (Chisholm and Tessier-Lavigne 1999; Semaphorin Nomenclature Committee 1999; Nakamura et al. 2000). Classes 3–7 are vertebrate semaphorins; class 3 semaphorins are secreted proteins, classes 4–6 are transmembrane proteins, and class 7 are GPI-anchored proteins (Nakamura et al. 2000). The semaphorins function is mediated by receptor complexes composed of neuropilins and plexins (Tamagnone and Comoglio 2000). It has been suggested that neuropilins are ligand-binding subunits and plexins are signaling subunits of class 3 semaphorin holoreceptor complexes (Raper 2000; Liu and Strittmatter 2001). In addition, it has been recently found that the cell adhesion molecule L1 is an essential component of the Sema3A receptor complex (Castellani et al. 2000). Recent studies suggest that Neuropilin-1 is the major receptor for the most studied Sema3A (He and Tessier-Lavigne 1997; Kolodkin et al. 1997), whereas Neuropilin-2 is the major receptor for Sema3F in the nervous system (Chen et al. 2000; Giger et al. 2000). Mutant Sema3A or Neuropilin-1 knock-out mice display very similar and striking defects in peripheral nerve projections (Behar et al. 1996; Kitsukawa et al. 1997; Taniguchi et al. 1997; Catalano et al. 1998), but many major CNS axon projections develop normally (Catalano et al. 1998). In contrast to Neuropilin-1 mutant mice, which die during the second half of gestation (Kitsukawa et al. 1997), many Neuropilin-2 mutant mice are viable into adulthood and thus suitable for the detailed analysis of axon guidance defects (Chen et al. 2000; Giger et al. 2000).

Sema3A acts as a chemorepellent for a number of axonal classes in vitro and in vivo: most rat cranial motor axons (Varela-Echavarria et al. 1997), cerebellar mossy fibres (Rabacchi et al. 1999), and neurites growing from pontine, cortical, and thalamic explants in vitro (Bagnard et al. 1998).

2.3 Slits Are Midline Repellents That Bind to Robo Receptors

Slits are secreted proteins involved in axon guidance and neuronal migration (Chisholm and Tessier-Lavigne 1999; Mueller 1999). Their receptor is the transmembrane protein Robo (Kidd et al. 1998a, b). Three *slit* genes (*slit1*, *slit2*, *slit3*) and three robo genes (*robo1*, *robo2*, *rig-1*) have been identified in mammals (Holmes et al. 1998; Itoh et al. 1998; Brose et al. 1999; Yuan et al. 1999). In *Drosophila*, Slits are repellent proteins made by midline glial cells (Kidd et al. 1998a, b, 1999), and play an important role in preventing commissural axons that have crossed the midline from recrossing the midline (Battye et al. 1999; Kidd et al. 1999). Commissural axons express low levels of Robo receptors on their surfaces prior to crossing the midline, due to the action of a negative regulator, the Commissureless protein (Kidd et al. 1998a, b). Thus, axons are able to cross the midline a single time. After crossing, however, the expression of Robo is dramatically upregulated, axons become highly responsive to the Slit repellent, and can no longer recross (Kidd et al. 1998a, b).

In vertebrates, commissural axons in the spinal cord are also responsive to the repellent activity created by floor plate cells; however, this activity is dependent on both Slits and Semaphorins (Zou et al. 2000). Slits are also expressed at the midline of the brain stem and the prosencephalon. For example, in the mouse, Slit proteins repel retinal axons in vivo, and cooperate to establish a corridor through which the axons are channeled and the optic chiasm forms in the ventral diencephalon (Erskine et al. 2000; Ringstedt et al. 2000; Plump et al. 2002). The role of Slits in the telencephalon is described below.

2.4 Ephrins and Eph Receptor Tyrosine Kinases Are Involved in Bidirectional Signaling

Ephrins are cell surface-associated ligands for Eph receptor tyrosine kinases implicated in repulsive axon guidance, cell migration, topographic mapping and angiogenesis (Flanagan and Vanderhaeghen 1998; Mellitzer et al. 2000; Knöll and Drescher 2002). In vertebrates, they comprise two major subclasses: EphA receptors and EphrinA ligands, and EphB receptors and EphrinB ligands (Mellitzer et al. 2000). Ephrin-

A1–A5 ligands are tethered to the cell surface by a GPI-anchor and bind EphA1–A8 receptors, whereas Ephrin-B1–B3 ligands are transmembrane proteins which posses a cytoplasmic portion and bind to EphB1–B6 and EphA4 receptors (Flanagan and Vanderhaeghen 1998; O'Leary and Wilkinson 1999; Klein 2001). In the Ephrin-Eph receptor signaling system, both classes of molecules function as receptors and as ligands. This phenomenon is described as bi-directional signaling and involves the mutual activation of particular signaling pathways in both of the interacting cells (Knöll and Drescher 2002). Whereas most receptor tyrosine kinases have soluble ligands, Ephrins must be membrane-bound in order to function properly (Knöll and Drescher 2002). The roles of Ephrins and Eph receptors in the developing telencephalon are described below.

3
Axonal Guidance Cues Display a Characteristic Spatio-Temporal Pattern of Expression in Embryonic and Fetal Zones of the Telencephalon

Fetal development of the telencephalon is characterized by continuous transformations and reorganization of the fetal cerebral wall which consists of transient, cytoarchitectonically defined compartments, the fetal zones (Rakic 1982; Kostovic and Rakic 1990; Kostovic et al. 1995, 2002; Kostovic and Judaš 2002). The cellular and fiber content of these zones is permanently changing; thus, fetal neuronal circuitry elements (afferent fibers, synapses, postsynaptic neurons) display transient patterns of areal, laminar, and modular organization (Kostovic et al. 1995). The fetal zones also display specific spatio-temporal patterns of expression of the ECM components and axonal guidance cues (Letourneau et al. 1994; Pearlman and Sheppard 1996; Kostovic et al. 2002). At least in the developing human brain, differences in the relative abundance and nature of ECM molecules enable the selective visualization of fetal zones (Kostovic et al. 2002). The transient subplate zone is a key compartment for transient fetal neuronal circuitry and competitive cellular interactions within the subplate zone are crucial for the areal specification of the cerebral cortex and the formation of cortical connectivity (Kostovic and Rakic 1990; Allendoerfer and Shatz 1994; Kostovic and Judaš 1995).

3.1 Proliferative Zones: The Ventricular Zone, the Subventricular Zone, and the Ganglionic Eminence

A number of axonal guidance cues are expressed in the neocortical ventricular zone (VZ) and the subventricular zone (SVZ), as well as in the ganglionic eminence (GE) which is a thickened portion of the VZ that serves as the primordium for the basal ganglia and some neocortical interneurons. Ephrin-A5, EphA3, EphA4 and EphA5 are expressed in both the VZ and SVZ (Mackarehtschian et al. 1999; Donoghue and Rakic 1999b) as well as in the GE and developing basal ganglia (Mackarehtschian et al. 1999). In the embryonic rodent telencephalon, the gradient of Ephrin-A5 in the VZ is one of the earliest markers for distinguishing cells residing in the anterior vs. posterior cortical neuroepithelium: Ephrin-A5 mRNA is most abundant in the anterior telencephalon, and expression decreases towards posterior regions (Mackarehtschian et al. 1999). Slit2 is expressed in the VZ that is present at and near the midline (Bagri et al. 2002). While the VZ expresses Sema3A, Sema4B, Sema5A and Sema5B (Adams et al. 1996; Giger et al. 1996; Bagnard et al. 1998; Catalano et al. 1998; Skaliora et al. 1998), the SVZ contains only Sema3A and Sema3C (Bagnard et al. 1998). It is interesting that the attractant Sema3C is produced by cells in the SVZ, whereas the repulsive Sema3A is produced by cells in both the VZ and SVZ, so that two partially overlapping antagonistic gradients are formed (Bagnard et al. 1998). This suggests that cortical axons are attracted to the intermediate zone by Sema3C, whereas Sema3A prevents the fibers from entering the SVZ/VZ neuroepithelium (Bagnard et al. 1998).

Two molecules involved in the pathfinding of corticofugal and thalamocortical axons through the nascent internal capsule display a characteristic expression in the ganglionic eminence. Netrin-1 is expressed in the GE of mice and rats at early developmental stages (Serafini et al. 1996; Livesey and Hunt 1997; Richards et al. 1997; Métin et al. 1997; Tuttle et al. 1999). The first cells to express Netrin-1 in the GE might be radial glial cells (Métin et al. 1997). In addition, Sema3A is also expressed in the ganglionic eminence (Skaliora et al. 1998). Interestingly, in the transient perireticular nucleus, Sema3A is expressed before birth, but is replaced by Sema5A after the birth (Skaliora et al. 1998).

In rodents, after birth, *slit1* is expressed in the medial globus pallidus, whereas *robo1* and *robo2* are expressed in the lateral globus pallidus (Marillat et al. 2002). In the caudate-putamen, *slit1* is restricted to scattered large neurons, *slit3* appears in the lateral part, and *slit2* is never detected in the striatum (Marillat et al. 2002). At P7, only Sema3A is present in the globus pallidus, whereas both Sema3A and Sema5B are present in the striatum (Skaliora et al. 1998). Moreover, the distribution of Sema3A is homogeneous, whereas Sema5B is distributed in discrete patches, suggesting that semaphorins are differentially involved in the development of striatal island and matrix compartments (Skaliora et al. 1998).

3.2 The Intermediate Zone: The Zone of Neuronal Migration and Axonal Growth

In the embryonic (E15-P0) rodent neocortex, the intermediate zone (IZ) is devoid of semaphorins (Skaliora et al. 1998) and Ephrin-A5 (Mackarehtschian et al. 1999), but contains Neuropilin-1 and Neuropilin-2 (Bagnard et al. 1998; Chédotal et al. 1998) and Robo2 (Marillat et al. 2002). In the same period, semaphorins are strongly expressed in the CP and SVZ-VZ, and thus appear to delineate the IZ as a territory through which axons are allowed to grow (Skaliora et al. 1998). Within this globally permissive zone, groups of fibers may be further separated based on their respective sensitivities to other diffusible factors, cell surface molecules or ECM cues (Sheppard et al. 1991; Bicknese et al. 1994; Miller et al. 1995; Pearlman and Sheppard 1996). In the white matter, semaphorins begin to be expressed only perinatally: Sema5B at P0, and Sema5A at P2; both increase further and are particularly pronounced in the middle of the dorsoventral extent of the white matter (Skaliora et al. 1998).

3.3 The Subplate Zone and Marginal Zone: Sites of the Earliest Synaptogenesis Between "Waiting" Axons and Transient Fetal Populations of Cells

Cajal-Retzius cells express the Netrin-1 receptor DCC (Keino-Masu et al. 1996) and the Slit receptor Robo2 (Nguyen-Ba-Charvet et al. 1999). In the rodent embryonic telencephalon, Slit3 is transiently present in the marginal zone from E15 to E18 (Marillat et al. 2002) The embryonic marginal zone also contains Sema3A which acts as a chemorepellent for axons of pyramidal neurons and as a chemoattractant for their apical dendrites (Polleux et al. 1998, 2000).

The subplate zone displays expression of Sema3A, Sema4B, Sema5A, and Sema5B (Skaliora et al. 1998), Robo2 (Marillat et al. 2002), Ephrin-

A3 and Ephrin-A5 (Mackarehtschian et al. 1999), EphA4, EphA5, EphA6, and EphA7 (Donaghue and Rakic 1999a, b; Mackarehtschian et al. 1999). Two of these guidance molecules have significantly restricted regional expression within the subplate zone, revealing that different regions of the early subplate zone are molecularly heterogeneous. First, the gradient of Ephrin-A5 is apparent in the rodent subplate zone and occurs in spatial opposition to a distinct gradient for the low-affinity NGF receptor p75 (Mackarehtschian et al. 1999). In contrast to the expression of the p75, which is mostly concentrated in the posterior half of the subplate, Ephrin-A5 is primarily expressed in the anterior subplate (Mackarehtschian et al. 1999). Second, Sema5A transcripts are restricted to the subplate of dorsomedial (but not ventrolateral) cortex; this constitutes the first instance of a regionally specific expression of a semaphorin gene along the tangential cortical dimension (Skaliora et al. 1998). Sema5A transcripts are restricted to the subplate zone of the dorsomedial cortex throughout its entire anteroposterior extent, suggesting that Sema5A expression defines the dorsomedial position across the cortical sheet rather than marking individual cortical areas. As this regionally specific distribution appears just before birth, it is probably not relevant to the selection of targets for thalamocortical afferents (which are already in place by then), but may be important for the routing of cortico-cortical or callosal projections which develop later (Skaliora et al. 1998).

3.4 The Cortical Plate: The Primordium of Future Cortical Layers II–VI

To date, the expression of guidance cues has been studied almost exclusively in the developing rodent cortex, in both embryonic (embryonic days E13–E20/E21) and early postnatal period (postnatal days P0–P14). The following guidance cues are found to be expressed in the cortical plate of the neocortex: Semaphorins Sema 3A, 3E, 3F, 4A, 4B and 5A (Catalano et al. 1998; Chédotal et al. 1998; Skaliora et al. 1998) and their receptor Neuropilin-1 (Kawakami et al. 1996; Chédotal et al. 1998); Slits1, 2 and 3 and their receptors Robo1 and Robo2 (Marillat et al. 2002; Whitford et al. 2002); Ephrins A5 and A3 and their Eph receptors EphA3–A7 (Castellani et al. 1998; Gao et al. 1998; Mackarehtschian et al. 1999). In addition, in the embryonic primate cerebral cortex, EphA3, A6 and A7 are also found to be expressed in the neocortical CP (Donoghue and Rakic 1999a, b; Šestan et al. 2000). Thus, there is a significant overlap in the expression of major guidance cue molecules in the cortical plate, with the notable exception of Netrin-1 which is never expressed in the neocortex (Keino-Masu et al. 1996; Chédotal et al. 1998; Braisted et al. 2000). However, these molecules are expressed at different developmental ages, in different cortical layers and cell types, and in uniformly wide or regionally restricted spatial patterns. Here, we focus on three important aspects of that differential expression: (1) the expression in the embryonic (compact and unlayered) cortical plate vs. the early postnatal cortical plate (when individual cortical layers sequentially differentiate in the inside-out pattern); (2) the postnatal expression in three major tangential cortical compartments: infragranular projection layers VI and V, granular thalamorecipient layer IV, and supragranular associative layers III and II; and (3) gradients and compartments of the restricted regional expression within the cortical plate and/or immature cortex.

Embryonic vs. Postnatal Cortex: In the embryonic neocortical plate, five semaphorins (Sema3A, 3E, 3F, 4A, and 4B) are expressed transiently and disappear shortly before the birth (Skaliora et al. 1998; Chédotal et al. 1998). The Neuropilin-1 (but not the Neuropilin-2) is expressed in the embryonic cortical plate concurrently with semaphorins (Kawakami et al. 1996, Chédotal et al. 1998). In the postnatal cortex, only Sema3A and Sema5A are expressed (Catalano et al. 1998; Skaliora et al. 1998). Ephrin-A3 and Ephrin-A5 are expressed in both the embryonic and postnatal rodent cortex (Mackarehtschian et al. 1999).

Infragranular, Granular, and Supragranular Layers. With respect to semaphorins, the main change in postnatal development is the appearance of Sema3A and Sema5A in the differentiating cortical layers, first in the infragranular and subsequently in the supragranular layers (Skaliora et al. 1998; Catalano et al. 1998). All cortical layers, with a notable exception of thalamorecipient layer IV (which never expresses any semaphorins!), express semaphorins as they become morphologically distinct. Infragranular layers express both Sema3A and 5A, whereas supragranular layers express only Sema3A (Skaliora et al. 1998). Similarly, only Slit1 is expressed in the embryonic cortical plate, whereas all three Slits are expressed in the postnatal cortex (Nguyen-Ba-Charvet et al. 1999; Marillat et al. 2002; Whitford et al. 2002). Robo1 and Robo2 are expressed in both the embryonic and postnatal cortex (Yuan et al. 1999; Marillat et al. 2002; Bagri et al. 2002). From P5 to adulthood, Robo2 is expressed in all cortical layers except in layer IV, while Robo1 becomes restricted to layers VI, V and the most superficial neurons of layer III; Slit1 is confined to layers VI and Va, Slit2 to layer V of the entire cortex and layers II/III of the cingulate cortex, whereas

Slit3 is expressed in layers II/III, Va and VI (Marillat et al. 2002; Whitford et al. 2002). According to some studies, Ephrin-A5, EphA3 and EphA5 are distributed across all cortical layers (Mackarehtschian et al. 1999), while others see expression of Ephrin-A5 mRNA limited to layer IV (Castellani et al. 1998). EphA4 expression is high in supragranular layers and low or absent in the deep layers (Mackarehtschian et al. 1999). In the embryonic primate cerebral cortex, EphA3 is expressed in developing layer IV, EphA6 in layer VI and the subplate zone, and EphA7 in developing layers V, VI and the subplate zone (Donoghue and Rakic 1999a, b).

Regional Expression. In the early postnatal cortex, Sema3A is distributed homogeneously throughout layer VI, but uniform Sema5A expression (in layers VI and V) is restricted mainly to the dorsomedial region, i.e., the cingulate cortex (Skaliora et al. 1998). More laterally, Sema5A is homogeneously distributed in layer V, but in layer VI it forms discrete columns; this patchy pattern is prominent in anterior and central regions (putative somatosensory cortex), but not in the posterior (putative visual) cortex (Skaliora et al. 1998). In addition, from P7 onwards, Sema3A selectively disappears from supragranular layers of the visual cortex (Skaliora et al. 1998). In the embryonic rodent cortex, Ephrin-A5 is expressed in a bi-directional gradient in the cortical plate, with highest levels in the somatosensory cortex; the abundance of the Ephrin-A5 message tapered off gradually towards anterior regions and more sharply towards posterior regions (Gao et al. 1998; Mackarehtschian et al. 1999). In the embryonic primate cerebral cortex, EphA6 appears to be a marker of both striate and extrastriate cortex early in development, while EphA3 is expressed only in the future extrastriate cortex (Donoghue and Rakic 1999a, b; šestan et al. 2000). Finally, Slit2 expression in postnatal cortical layers II/III is confined to the cingulate cortex (Marillat et al. 2002; Whitford et al. 2002).

4 Guidance Cues Have Complementary and/or Overlapping Roles in the Establishment of Specific Axonal Pathways

4.1 Commisural Pathways: The Corpus Callosum, the Anterior Commisure, and the Hippocampal Commisure

Studies in the embryonic cat, rat, and mouse brain have suggested that early midline astroglia provide the substrates for the formation of commissural fiber tracts in the telencephalon, i.e., the corpus callosum (CC), the anterior commissure (AC), and the hippocampal commissure (HC; Silver and Ogawa 1983; Silver et al. 1993). Several guidance molecules are expressed in telencephalic regions populated by these midline glia: Netrin-1 (Serafini et al. 1996; Livesey and Hunt 1997), Slits (Li et al. 1999; Nguyen-Ba-Charvet et al. 1999, 2002; Wu et al. 1999; Shu and Richards 2001; Bagri et al. 2002), EphB3 (Orioli et al. 1996; Bertuzzi et al. 1999), and Vax1 (Hallonet et al. 1998; Bertuzzi et al. 1999).

Netrin-1-expressing midline cells partially envelop these commissural axons (Serafini et al. 1996), and Netrin-1 is also expressed along the path of the fimbria (Serafini et al. 1996; Keino-Masu et al. 1996; Chédotal et al. 1998). Slit2 is expressed in two glial populations (glial wedge and indusium griseum) in the region where the corpus callosum forms (Shu and Richards 2001; Bagri et al. 2002). The callosal axons extend into a narrow pathway that forms between the two populations (Bagri et al. 2002). Slit2 derived from the glial wedge appears to prevent callosal axons from entering the septum, whereas Slit2 from the indusium griseum may prevent the axons from traveling dorsally back into the ipsilateral cortex, instead channeling them across the midline to the contralateral side; Slit proteins appear to regulate midline crossing by channeling axons into an appropriate crossing site, rather than by regulating crossing or recrossing at the midline per se (Bagri et al. 2002). Neuropilin-1 is expressed in cortical neurons that give rise to the corpus callosum (Kawakami et al. 1996; Polleux et al. 1998), as well as in the fibers of the corpus callosum (Giger et al. 2000). Neuropilin-2 is expressed in fibers of the anterior commissure (Giger et al. 2000; Chen et al. 2000). In addition, AC axons express Ephrin-B ligands, while the EphB2 and EphB3 receptors are found in the vicinity of the pathway, in the ventral region which the axons normally avoid (Henkemeyer et al. 1996; Orioli et al. 1996). Finally, axons of all three commissures express the cell adhesion molecule TAG-1, whereas callosal and hippocampal commissural neurons express both TAG-1 and L1 (Wolfer et al. 1994; Fujimori et al. 2000).

In mice, the targeted disruption of genes encoding various axonal guidance molecules results in loss of one, two, or all three commissures in the telencephalon. When commissures are absent (although midline fusion was normal), selective fasciculation between commissural axons is usually inhibited, and axons become tangled bilaterally into "Probst's bundles", i.e., swirls of axons on either side of the midline. All three commissures are lost in Netrin-1-deficient homozygous mice (Keino-Masu et al. 1996; Serafini et al. 1996), Vax1-deficient mice (Bertuzzi et al. 1999), and growth-associated protein-43 (GAP-43) mutant mice

(Shen et al. 2002). However, in Netrin-1 heterozygotes, the only obvious defect seen is the presence of cell bodies within the hippocampal commissure that appear largely absent in wild-type animals (Serafini et al. 1996). Similarly, in heterozygote GAP-43 mutant mice the anterior commissure was unaffected (Shen et al. 2002).

In Neuropilin-2 mutant mice, there is a dramatic loss of the anterior commissure, but there are no defects in the corpus callosum and the hippocampal commissure (Giger et al. 2000; Chen et al. 2000). Targeted disruption of the *ephB2* gene results only in the loss of the posterior part of the anterior commissure (Henkemeyer et al. 1996), whereas targeted disruption of *ephB3* affects formation of the corpus callosum (Orioli et al. 1996). Double homozygotes with disruptions of both *ephB2* and *ephB3* show synergistic midline defects in both the AC and the CC, providing direct evidence for functional redundancy among the Eph receptors (Orioli et al. 1996).

In vitro, Slit2 can repel outgrowing cortical axons (Shu and Richards 2001). In Slit2 mutant mice, the size of the corpus callosum is severely reduced, but both the anterior commissure and the hippocampal commissure appear grossly normal (Bagri et al. 2002). Finally, the corpus callosum is impaired in L1-deficient mice (Demyanenko et al. 1999) and in certain patients with L1 mutation (Kamiguchi et al. 1998). As previously mentioned, the L1 is a component of the Sema3A receptor complex (Castellani et al. 2000). Thus, callosal projections may require Sema3A-L1 functional interaction for their guidance (Castellani et al. 2000).

4.2 Thalamocortical and Corticothalamic Pathways

To navigate from the thalamus to the neocortex, thalamocortical axons use multiple molecular guidance cues secreted by tissues lying along their path. Thalamocortical axons use the axons of neurons situated in the internal capsule that project into the dorsal thalamus as a scaffold to enter the internal capsule in hamster (Métin and Godement 1996), rat (Molnar et al. 1998a, b), and mouse (Braisted et al. 1999; Tuttle et al. 1999). In addition, pioneer axons of subplate cells might provide a scaffold for the later growing cortical plate neurons (McConnell et al. 1989). Growing thalamocortical axons encounter subplate axons in the internal capsule (Molnar et al. 1998a, b; Auladell et al. 2000), and then turn tangentially in the intermediate zone. Indeed, the junction between the internal capsule and the intermediate zone has recently been identified as an important decision-making region for growing thalamic axons

(Skaliora et al. 2000). When thalamic axons reach the IZ, they slow down, often coming to a complete stop for up to several hours, and their growth cones become larger and more complex. During pauses there are continuous extensions and retractions of filopodia and/or side branches. When advance resumes, it is often towards a different direction (Skaliora et al. 2000). These findings strongly suggest that at the transition from the internal capsule to the intermediate zone, thalamocortical axons are receiving and integrating a novel set of cues to make the next pathfinding decision (Skaliora et al. 2000).

According to the "handshake hypothesis", the early thalamic axons use subplate axons as a scaffold on their way to the cortex and that, in fact, subplate and thalamocortical axons might fasciculate and thereby guide each other towards their final destination (Molnar and Blakemore 1995; Molnar et al. 1998a, b). For example, limbic thalamic axons form fascicles with limbic cortical axons, and that fasciculation is mediated by the LAMP, a protein expressed on both limbic thalamic and limbic cortical axons (Mann et al. 1998). However, it was shown recently that axonal surface molecules act in combination with Sema3A to increase the rate of homotypic fasciculations and, at the same time amplify the segregation between cortical and thalamic axons during the establishment of corticothalamic projections (Bagnard et al. 2001). These results suggest that homotypic axon fasciculation in combination with heterotypic fiber repulsion leads to segregated tracts for the majority of thalamic and cortical axons (Bagnard et al. 2001). In addition, in COUP-TF1 mutant mice, thalamic axons manage to enter the internal capsule, but very few thalamic fibers are able to project out of the internal capsule to reach the cortex (Zhou et al. 1999). At the same time, subplate cells in COUP-TF1 mutants project normally towards the thalamus. Thus, it seems that the subplate scaffold alone is not sufficient to guide thalamic axons to their final cortical target (Zhou et al. 1999).

Recently, two molecules have emerged as key guidance cues for attracting both corticothalamic and thalamocortical axons to the nascent internal capsule within the ganglionic eminence (Netrin-1), and for confining their trajectories to the internal capsule and intermediate zone (Sema3A).

A population of Netrin-1-expressing cells in the ventral telencephalon and the ganglionic eminence attracts both corticofugal (Métin et al. 1997; Richards et al. 1997) and thalamocortical axons (Braisted et al. 2000). In *netrin-1* null mutant mice, thalamocortical projection through the ventral telencephalon is disorganized, their pathway is abnormally restricted, and fewer dorsal thalamic axons reach the cortex (Braisted et al. 2000). Netrin-1 is also lacking in *mash-1*-deficient mice and, as a consequence, the thalamocortical projection is dramatically altered in Mash-1 null mutant mice (Tuttle et al. 1999). In Mash-1-deficient mice, thalamocortical axons do not proceed into the internal capsule, but instead form a dense knot reminiscent of the Probst bundles between the dorsal and ventral thalamus (Tuttle et al. 1999). Finally, the Netrin-1 acts in vitro as an attractant and growth promoter for dorsal thalamic axons (Braisted et al. 2000), and Netrin-1 receptors DCC and neogenin are expressed in the dorsal thalamus (Braisted et al. 2000). It should be noted that the ganglionic eminence and developing striatum express at least one other chemoattractant for axons from cortical explants – Sema3C (Bagnard et al. 1998).

On the other hand, the hypothalamus is repulsive for thalamocortical axons (Braisted et al. 1999). The Sema3A is expressed in the developing hypothalamus (Zhou et al. 1997) and both Sema3A and Sema5A are expressed in the ganglionic eminence and the developing striatum (Giger et al. 1996; Skaliora et al. 1998; Bagnard et al. 2001). Sema3A is also expressed in the thalamic reticular nucleus (Skaliora et al. 1998) and in the telencephalic ventricular zone (Catalano et al. 1998; Skaliora et al. 1998). Sema3A can collapse growth cones of dorsal thalamic axons (Bagnard et al. 1998), and has been shown to repel thalamic axons in culture (Bagnard et al. 1998; Polleux et al. 1998). Although Sema3A is strongly expressed along the thalamocortical pathway, Neuropilin-1 and Neuropilin-2 do not appear to be present in the embryonic rodent dorsal thalamus (Kawakami et al. 1996; Chen et al. 1997) and thalamocortical axon pathfinding seems to be normal in mice deficient for Sema3A (Behar et al. 1996; Catalano et al. 1998). Both thalamocortical and corticothalamic fibers in vivo still succeed in entering the Sema3A-positive developing striatum, and crossing it in large fascicles (Skaliora et al. 1998). Therefore, it has been proposed that the Sema3A may be acting as a partial repellent, allowing growth, but preventing branching or synapse formation (Skaliora et al. 1998). Thus, Sema3A could help to solve the general problem that presents itself when outgrowing axons need to traverse cell-dense regions without engaging in branching and synapse formation. Interestingly, thalamic axons branch selectively within the cortical layer IV, the only cortical layer that does not express any of the semaphorin genes (Skaliora et al. 1998). Thus, Sema3A could provide a surround repulsion that prevents branching and synapse formation in inappropriate regions (Skaliora et al. 1998).

In the embryonic rhesus monkey, prospective visual cortical areas and corresponding thalamic nuclei can be defined by complementary expression of Ephrins and Eph receptors, prior to obvious cytoarchitectonic differentiation within the cortical plate and before the establishment of reciprocal connections between the cortex and thalamus (Šestan et al. 2000). Thus, ephrins and EphA family members may be involved in both compartment formation and axon guidance within the visual thalamocortical system (Šestan et al. 2000). Similarly, in the developing rodent thalamus, Ephrin-A5 and EphA3-A5 show distinct patterns of expression (Mackarehtschian et al. 1999). The Ephrin-A5 inhibits axon growth by limbic neurons in both the thalamus and cortex, which may prevent the inappropriate innervation of the primary sensorimotor area by these neurons (Gao et al. 1998). Gene targeting experiments in mice have also demonstrated that Ephrin-A5 is required for the topographic organization of visual inputs to the thalamus (Feldheim et al. 1998). It has been recently suggested that thalamic axons might identify their target area by using the graded expression in the subplate of an axon pathfinding cue such as Ephrin-A5 (Mackarehtschian et al. 1999). However, mRNAs for Robo1, Robo2 and all three Slits are also expressed in dorsal thalamus (Bagri et al. 2002; Marillat et al. 2002), and thalamocortical axons avoid three regions with high Slit mRNA expression: the midline, the preoptic area and hypothalamus, and the proliferative neuroepithelium of the ganglionic eminence (Bagri et al. 2002). Moreover, in *slit2* mutants, and even more prominently in *slit1/slit2* double mutants, thalamocortical fibres of the inferior thalamic radiation turned ventrally and caudally to enter the hypothalamus, and many of these fibers then turn medially and approach the midline (Bagri et al. 2002). Finally, Sema6A has also been shown to play a role in guiding thalamic axons in vivo (Leighton et al. 2001).

4.3 Corticofugal Pathways

Semaphorins seem to be involved in guiding cortical axons (Giger et al. 1996; Skaliora et al. 1998). It has recently been shown that Sema3C can attract cortical axons, whereas Sema3A repels cortical axons and inhibits axonal branching (Bagnard et al. 1998). In addition, Sema3A can convert a permissive substratum into a substratum that is repellent for cortical axons, thereby preventing these fibers from entering specific regions delineated by this molecule (Bagnard et al. 1998). In mice, layer V pyramidal neurons normally express high levels of Sema3A, especially after the birth (Catalano et al. 1998). In Sema3A knockout mice, cortical neuropil is reduced and axons of pyramidal neurons are poorly oriented towards the ventricular surface (Behar et al. 1996). Moreover, Sema3A repels cortical projection axons in vitro (Polleux et al. 1998), and axons of pyramidal cells orient less precisely towards the ventricular surface when plated onto cortical slices made from Sema3A knockout mice (Polleux et al. 2000). These findings suggest that Sema3A expression in the cortical marginal zone helps to orient the initial outgrowth of pyramidal axons by repelling them towards the ventricular surface (Polleux et al. 1998, 2000).

After leaving the cortical plate and the subplate zone, corticofugal axons enter the IZ where they express both Neuropilin-1 and Neuropilin-2 mRNAs (Bagnard et al. 1998; Chédotal et al. 1998). Growing corticofugal axons might require attractant signals that drive them deep into the IZ, in combination with repulsive molecules that prevent these fibers from growing into the SVZ/VZ, as well as another attractant signal that steers them towards the internal capsule. As previously mentioned, the IZ is devoid of semaphorins, whereas semaphorins are strongly expressed in the CP and SVZ/VZ (Skaliora et al. 1998; Bagnard et al. 1998). Thus, semaphorins appear to delineate the IZ as a territory through which axons are allowed to grow (Skaliora et al. 1998). Moreover, the attractant Sema3C is produced by cells in the SVZ, whereas the repulsive Sema3A is produced within both the VZ and SVZ, so that two partially overlapping antagonistic gradients are formed (Bagnard et al. 1998). This suggests that cortical axons are attracted towards the intermediate zone by Sema3C, whereas Sema3A prevents the fibers from entering the SVZ/VZ neuroepithelium (Bagnard et al. 1998). Finally, in order to pass from the IZ into the internal capsule, corticofugal axons are first attracted to an intermediate target, the ganglionic eminence (Métin and Godement 1996). Recent evidence suggests that Netrin-1 mediates these effects (Métin et al. 1997; Richards et al. 1997; Braisted et al. 2000). Netrin-1 is expressed in the GE of mice and rats at early developmental stages (Serafini et al. 1996; Livesey and Hunt 1997; Richards et al. 1997; Métin et al. 1997; Tuttle et al. 1999). The first cells to express Netrin-1 in the GE might be radial glial cells (Métin et al. 1997). Interestingly, corticofugal axons extending into the lateral ganglionic eminence never enter the ventricular neuroepithelium, even though it also strongly expresses Netrin-1 (Métin et al. 1997). This suggests that the VZ is an unfavorable environment for cortical axon outgrowth, and that its inhibitory action cannot be overridden by Netrin-1 (Métin et al. 1997). Sema3A is expressed in both the VZ and the ganglionic eminence (Skaliora et al. 1998). However, corticofugal fibers avoid the GE as well as the hypothalamic region because, in addition to semaphorins, these regions also express Slits (Bagri et al. 2002). In *slit2* mutant mice, cortical axons in the internal capsule are displaced ventrally and thick fascicles of axons are found in an abnormal position close to the ventral surface of the basal telencephalon. In addition, some cortical axons inappropriately approach the midline in *slit2* mutants (Bagri et al. 2002).

A wide spectrum of neurological disorders, collectively referred to as CRASH syndrome, has been attributed to mutations in the X-linked gene encoding the neural cell adhesion molecule L1 (Kamiguchi et al. 1998). A common feature of these disorders is the abnormal development of the corticospinal tract, and subsequent analysis of L1-deficient mice has confirmed an essential role for L1 in corticospinal tract development (Dahme et al. 1997; Cohen et al. 1998). Surprisingly, the L1 is an essential component of the Sema3A receptor complex, and L1 mutations may disrupt Sema3A signaling in the growth cone, leading to guidance errors (Castellani et al. 2000). The region of pyramidal decussation coincides with an abrupt upregulation of ventral Sema3A expression, which deflects cortical axons away from the ventral spinal cord, to cross the midline and to enter the contralateral dorsal column (Castellani et al. 2000). In addition, coculture assays have demonstrated that Sema3A secreted from the ventral spinal cord repels cortical axons from wild-type, but not from L1-deficient mice (Castellani et al. 2000).

Finally, it has been recently demonstrated that the Ephrin-B3, expressed at the spinal cord midline, prevents EphA4-expressing corticospinal tract axons from aberrantly crossing to the other side (Kullander et al. 2001; Leighton et al. 2001; Yokoyama et al. 2001).

4.4
Ascending Monoaminergic Pathways

Recently, two molecular cues have been implicated in the guidance of growing serotonergic and dopaminergic fibers towards the cortex: Slits (Bagri et al. 2002), and growth-associated protein-43, GAP-43 (Donovan et al. 2002). In *slit2* mutant mice, monoaminergic fibers in the medial forebrain bundle are displaced ventrally as they course through the diencephalon, whereas in *slit1/slit2* double mutants, the bundle is commonly split into two components and numerous fibers descend ventrally into the hypothalamus, approaching the midline (Bagri et al. 2002). Although many serotonergic and dopaminergic fibers enter the telencephalon normally, in *slit1/slit2* double mutants a significant percentage of fibers abnormally cross the midline in the basal telencephalon (Bagri et al. 2002). The GAP-43 is a growth-associated protein expressed on growing serotonergic axons; a nearly complete failure of serotonergic axons to innervate the cortex and hippocampus is described in mice lacking GAP-43 (Donovan et al. 2002). However, the density of serotonergic axons is normal in the striatum, septum, and amygdala and is even dramatically higher than normal in the thalamus of GAP-43 mutant mice (Donovan et al. 2002).

4.5 Guidance Cues in Developing Afferent, Efferent and Intrinsic Pathways of the Hippocampal Formation

Recent studies suggest that class 3 semaphorin-neuropilin signaling is likely to play an important role in the stereotypical generation of correct hippocampal afferent and efferent pathways (Chédotal et al. 1998; Steup et al. 1999). Axons from CA1, CA3 and dentate gyrus mouse embryo explants are selectively repelled by the entorhinal cortex and neocortex; entorhinal axons are repelled by embryonic neocortical tissue, whereas neocortical axonal outgrowth is unaffected by nearby hippocampal explants (Chédotal et al. 1998). Current research is focused on possible roles of two pairs of semaphorins and their receptors in these events: Sema3A-Neuropilin-1, and Sema3F-Neuropilin-2 (Chen et al. 1997, 2000; Chédotal et al. 1998; Giger et al. 1998, 2000).

Sema3F mRNA is uniformly expressed in cortical plate cells and their axons in the hippocampal fields CA1 and CA3, the entorhinal cortex, and the adjacent neocortex, but not in the dentate gyrus (Chédotal et al. 1998). Sema3A mRNA was widely expressed in the neocortical and entorhinal cortical plate, but virtually absent from the hippocampus (Chédotal et al. 1998). Neuropilin-1 is expressed in the CA3 (Giger et al. 2000; Chen et al. 2000). Neuropilin-2 is expressed in the septum, CA3, CA1, dentate gyrus, and fimbria-fornix system (Chen et al. 1997, 2000; Chédotal et al. 1998; Giger et al. 1998, 2000). Sema3A and Sema3F strongly repel CA1, CA3 and dentate gyrus axons, whereas entorhinal axons are only repelled by Sema3A (Chédotal et al. 1998; Steup et al. 1999). In *neuropilin-2* mutants, the fornix is present, but appears less tightly bundled (Giger et al. 2000). The ventral hippocampal commissure appears normal in all mutants, suggesting that hippocampal commissure does not depend on Sema3F-Neuropilin-2 interactions (Giger et al. 2000). Defects of *neuropilin-2* mutants include the overshooting of dentate gyrus mossy fibers towards their targets in the CA3 field (Giger et al. 2000; Chen et al. 2000). All of these findings suggest that the correct termination of dentate mossy fibers on neurons in the CA3 field results from Sema3F-mediated repulsion (Chédotal et al. 1998; Giger et al. 2000). The high expression of Sema3F in the hippocampal plate may direct and confine pyramidal cell axons to neighboring layers, such as the intermediate zone and stratum radiatum (Chédotal et al. 1998).

Slit and Robo proteins have also been implicated in the development of hippocampal connectivity (Nguyen-Ba-Charvet et al. 1999). Slit1 and Slit2, as well as Robo1 and Robo2 are all expressed in both CA1 and CA3 fields (Nguyen-Ba-Charvet et al. 1999; Marillat et al. 2002). The subiculum contains Slit1 and Robo1 which is later replaced by Robo2; the presubiculum displays only Slit2, whereas the entorhinal cortex and dentate gyrus express only Robo1 (Marillat et al. 2002). Slit2 is a chemorepellent molecule for dentate gyrus axons and may have a function in pushing noncommissural axons away from the midline of the forebrain (Nguyen-Ba-Charvet et al. 1999).

Finally, complementary gradients of receptor and ligand expression and binding were found in the hippocamposeptal system, with Ephrins A2, A3 and A5 expressed in the septum and EphA5 in the hippocampus (Gao et al. 1996; Zhang et al. 1996). In vitro, Ephrin-A2 can inhibit neurite outgrowth from explants of the topographically appropriate medial, but not lateral hippocampus, suggesting that Ephrin-A2 could function as a topographic mapping molecule in the hippocamposeptal map (Gao et al. 1996).

4.6
Guidance Cues in the Developing Olfactory System

In the rodent brain, axons from olfactory receptor neurons in the olfactory epithelium project ipsilaterally to mitral and tufted cells of the main olfactory bulb. Axons of olfactory bulb neurons project ipsilaterally to higher olfactory centers in the telencephalon and form the lateral olfactory tract (LOT; Zou et al. 2001; Nguyen-Ba-Charvet et al. 2002). The pathfinding of LOT axons in the telencephalon is controlled by a combination of short-range and long-range cues (Pini 1993; Cloutier et al. 2002; Nguyen-Ba-Charvet et al. 2002). For example, diffusible factors that repel LOT axons in vitro are secreted by the septum (Pini 1993; Hu and Rutishauser 1996) and the neocortex (Chédotal et al. 1998; Giger et al. 1998; De Castro et al. 1999). On the other hand, the mesenchyme precursor of the frontal bone which is juxtaposed to the pathway followed by LOT axons serves as a source of a diffusible chemoattractant for these axons (De Castro et al. 1999).

The cells of the embryonic olfactory bulb express Sema3A (Schwarting et al. 2000; Cloutier et al. 2002) and Slit1 in a subset of mitral cells (Nguyen-Ba-Charvet et al. 1999). Growing LOT axons express Neuropilin-1 (Chen et al. 1997) and Robo2 (Nguyen-Ba-Charvet et al. 1999). In vitro, LOT axons are attracted by Sema3B (De Castro et al. 1999), but repelled by Sema3A (Kobayashi et al. 1997; Schwarting et al. 2000),

Sema3F (Chédotal et al. 1998; Giger et al. 1998; De Castro et al. 1999), and by the Slit2 (Li et al. 1999; Nguyen-Ba-Charvet et al. 1999; Chen et al. 2001). The chemoattractant Sema3B is expressed in the frontal bone primordium; this finding could explain why LOT axons grow preferentially in the most superficial zone of the forebrain (De Castro et al. 1999). If confirmed, these results would represent the first evidence for a presumptive influence of nonneuronal peripheral tissue on axon tract formation and orientation in the vertebrate brain (De Castro et al. 1999).

The repellent Sema3F is produced by the adjacent neocortex (Chédotal et al. 1998; Giger et al. 1998; De Castro et al. 1999), and Slit repellents are produced by the septum (Nguyen-Ba-Charvet et al 1999, 2002; Marillat et al. 2002).

In mice lacking Sema3A, many LOT axons display aberrant trajectories and connections are made in inappropriate target regions (Schwarting et al. 2000). However, the LOT develops normally in both Neuropilin-1 deficient mice (Kitsukawa et al. 1997) and Neuropilin-2 mutant mice (Chen et al. 2000). On the other hand, Neuropilin-2 is strongly expressed in a subset of vomeronasal sensory neurons and in their targets in the accessory olfactory bulb (Kawakami et al. 1996; Chen et al. 1997; Giger et al. 1998). In Neuropilin-2 mutant mice the vomeronasal nerve is defasciculated and a subset of vomeronasal neurons ectopically innervate the main olfactory bulb (Cloutier et al. 2002).

Finally, the LOT develops normally in Slit1 or Slit2 single-deficient mice, but is completely disorganized in Slit1/Slit2 double-deficient embryos, with many axons reaching the midline and entering the septum region (Nguyen-Ba-Charvet et al. 2002). Therefore, the septum chemorepellent is a combination of Slit1 and Slit2 and these molecules play a significant role in LOT axon guidance in vivo.

5 Early Genes Involved in Regionalization and Areal Specification of the Cerebral Cortex Can Regulate the Pattern of Expression of Axonal Guidance Cues

Recent studies provide experimental evidence for the involvement of regulatory genes in the control of the patterned expression of axon guidance molecules. Of course, axon guidance molecules may become distributed outside of their domains of gene expression if they are present on axons extended outside of the domain by expressing neurons (Tuttle et al. 1999).

In vertebrates, *netrin-1a* expression is regulated by hedgehog signaling, and ectopic expression of hedgehogs induces ectopic *netrin-1a* expression in the neural tube (Lauderdale et al. 1998). Similarly, ectopic expression of the engrailed genes, *en-1* and *en-2*, in the chick optic tectum induces the expression of the genes that encode Ephrin-A2 and Ephrin-A5 (Logan et al. 1996; Shigetani et al. 1997). In Mash-1-deficient mice, the thalamocortical projection is dramatically altered because the Netrin-1 expression is lacking in a crucial decision-making point area for these axons in the ventral telencephalon (Tuttle et al. 1999). Interestingly, disregulation of genes encoding competing positive and negative guidance cues underlies at least some of the axon guidance defects. For example, the homeodomain protein Vax1 regulates expression of the axon guidance cues of a set of anterior midline cells that orchestrate axon trajectories in the developing mammalian forebrain (Bertuzzi et al. 1999). In *vax1* mutant mice, the expression of the chemoattractant Netrin-1 by midline cells is severely affected, but the expression of the chemorepellent Slit1 is not affected (Bertuzzi et al. 1999). Therefore, it has been suggested that the loss of Vax1 shifts the normally balanced presentation of attractive and repellent guidance cues at the anterior midline in favor of the latter (Bertuzii et al. 1999). That is, axons now confront a sharp and discontinuous repulsive barrier where chemorepellent activities of molecules like Slit-1 are no longer modulated by the chemoattractant activities of cues such as Netrin-1 (Bertuzzi et al. 1999).

6 Discussion and Conclusions

The pathfinding of most axons is dependent upon a complex combination of secreted and membrane-bound growth cone attractants and repellents (Mueller 1999). Moreover, many guidance cues are bifunctional (Song and Poo 2001). For example, the levels of second messengers within the growth cone can influence whether a growth cone responds to a particular cue with attraction or repulsion (Ming et al. 1997; Hong et al. 2000). In addition, ECM components which promote axon outgrowth can modify the response of growth cones to axonal guidance cues (Hopker et al. 1999). This raises the possibility that the nature of the substrate might also influence interactions between growing axons.

Recent findings have also led to a re-examination of the classical notion of "diffusible" guidance cues. For example, it is likely that all vertebrate semaphorins are tightly localized on or near the cells in which they are made (Raper 2000). Thus, depending on the composition and spatial arrangement of the extracellular environment, secreted semaphorins might not be distributed uniformly and, instead, may form gradients or sharp boundaries in vivo (Bagnard et al. 1998). This suggests that, depending on the spatial distribution, a given molecule can have very different effects on a specific population of cortical neurons (Bagnard et al. 1998). The semaphorins are obviously bifunctional molecules whose effects depend on their spatial distribution (Bagnard et al. 1998; Skaliora et al. 1998). Therefore, it was suggested that, at least in the developing cortex, molecules that influence axon behavior should not be classified according to their action on a single cell population or in a single way (Bagnard et al. 1998, 2000). Rather, as proposed by Bolz and Castellani (1997), such molecules should be called collectively "wiring molecules": although they can function in multiple and alternative ways, they serve as signals for assembling the intricate network of cortical connections.

Another important question is how axons, steered towards intermediate targets by chemoattractants, manage to grow past those intermediate targets. It has been recently demonstrated that stabilized gradients of increasing semaphorin concentrations elicit stereotypical responses from cortical growth cones, independent of the absolute concentration and slope of these gradients (Bagnard et al. 2000). Thus, growth cone guidance by gradients of chemotropic molecules is robust and reproducible, because it is primarily independent of the exact dimensions of the gradients – instead, growth cones are very sensitive to the direction or "sign" of gradients (Bagnard et al. 2000). The capacity of growth cones to ignore the decreasing concentration of chemoattractants may represent one of the key strategies which allows axons to grow past intermediate targets (Bagnard et al. 2000).

The secreted semaphorins most often appear to act as repellents rather than as attractants (Raper 2000). In addition, the semaphorin activity in both invertebrates and vertebrates seems to affect axon fasciculation (Raper 2000). The distribution of semaphorins also displays interesting regional features (Skaliora et al. 1998): Sema5A is always present in all neuroepithelia throughout the brain, whereas Sema4A is never found in proliferative tissue; Sema4A is never found in the basal ganglia or the thalamus, and seems to be expressed only in structures derived from the dorsal telencephalon; Sema4B is never expressed in either the thalamus or the hippocampus, and is confined to the neuroepithelium and the cortical plate. In addition, Sema5A and 5B are the only two semaphorins ever seen in the fiber tracts of cortical white matter, e.g., the corpus callosum and the internal capsule (Skaliora et al. 1998). On the other hand, in cortical layer IV none of semaphorin genes are ever expressed (Skaliora et al. 1998). Even within the same cortical layer (layer VI), Sema3A and Sema5A are expressed in different cell populations (Skaliora et al. 1998). Similar observations are published for other guidance molecules. In the embryonic primate telencephalon, EphA3 and AphA7 are expressed in both the neocortex and the striatum, whereas the EphA6 expression is restricted to the neocortex (Donoghue and Rakic 1999a, b). While most neurons express at least one *robo* and one *slit* gene during their development, no Robos were detected in the septum and in neocortical layer IV (Marillat et al. 2002).

As previously described, an important feature of some axonal guidance molecules is their regionally restricted expression in certain fetal zones during the period critical for a defined histogenetic event which predominantly enfolds in a given fetal zone. For example, Sema5A transcripts are restricted to the subplate zone of the dorsomedial cortex and may be important for the routing of cortico-cortical or callosal projections (Skaliora et al. 1998). The gradient of Ephrin-A5 is apparent in the rodent subplate zone and may be significant for the sequential ingrowth of thalamocortical afferents to proper cortical areas (Mackarehtschian et al. 1999). In addition, some guidance cues seem to be involved in the regional and/or areal specification of the cortical plate. For example, Sema5A (Skaliora et al. 1998) and Ephrin-A5 (Gao et al. 1998; Mackarehtschian et al. 1999) may be involved in the specification of the rodent somatosensory cortex, whereas in the embryonic primate cortex EphA6 and EphA3 serve as early markers of striate and extrastriate cortical areas (Donoghue and Rakic 1999a, b). In the embryonic rodent brain, Slit3 is found almost exclusively in the hippocampal formation, and *slit3* is one of the first genes selectively expressed in the embryonic hippocampal formation (Marillat et al. 2002). Finally, Slit2 expression in postnatal cortical layers II/III is confined to the cingulate cortex (Marillat et al. 2002; Whitford et al. 2002).

Several generalizations about the roles of specific guidance cues in the developing telencephalon have emerged to date. With respect to semaphorins, three main changes in postnatal cortical development are: (1) the gradual addition of semaphorin transcripts in the subplate zone; (2) the onset of Sema5A and 5B expression in the cortical white matter, and (3) the appearance of Sema3A and 5A in the differentiating cortical layers, first in the infragranular and subsequently in the supragranular layers (Skaliora et al. 1998). According to Skaliora et al. (1998), semaphorins are in the position to: (1) prevent axon extension into the ventricular neuroepithelium throughout the brain; (2) confer nonpermissive properties to the embryonic cortical plate, hence regulating the radial invasion of corticopetal afferents; (3) confine axonal extension to the intermediate and the subplate zone; (4) maintain the fasciculated state of thalamocortical and corticothalamic axons, and prevent them from branching while they grow through the striatum; and (5) restrict the terminal arborizations of thalamic afferents to layer IV. In addition, Sema3A is also strongly expressed in structures of the olfactory-hippocampal pathway and the motor system in the adult rat and human brain (Giger et al. 1998). Thus, Sema3A might be also involved in restricting structural changes that occur in the wiring of the intact adult brain (Giger et al. 1998). The Robo2 expression is tightly (but not exclusively) associated to developing olfactory, visual, auditory, and somatosensory systems during the critical period of axon outgrowth and pathfinding (Marillat et al. 2002).

Slit proteins in the telencephalon appear to contribute to: (1) the maintenance of dorsal position by prevention of axonal growth into ventral regions; (2) the prevention of axonal extension toward and across the midline, and (3) the channeling of axons toward particular regions (Bagri et al. 2002). In most regions of the rat brain, slit and robo are expressed in a complementary pattern, and their expression is up-regulated postnatally, suggesting that they could also be involved in synaptic plasticity (Marillat et al. 2002).

Many axonal guidance molecules have important additional roles in the developing brain. For example, Netrin-1 (in combination with its receptor DCC) functions as a midline chemoattractant not just for axons, but for the directed circumferential migration of basilar pontine neurons (Serafini et al. 1996; Fazeli et al. 1997; Yee et al. 1999) and inferior olivary neurons (Bloch-Gallego et al. 1999). Slits have been shown to repel several classes of tangentially migrating neurons in the telencephalon: neurons migrating from the anterior subventricular zone to the olfactory bulb (Chen et al. 2001), precursors of interneurons in the postnatal olfactory bulb (Hu 1999; Wu et al. 1999), and GABAergic neurons that migrate from the striatal primordium to the neocortex (Zhu et al. 1999). Interestingly, the repulsive activity of Slit on migrating subventricular zone neuroblasts requires cell-surface heparan sulfate proteoglycan (Hu 2001). Slit2 can also stimulate axonal elongation and branch formation of sensory axons in the rodent dorsal root ganglia (Wang et al. 1999) and central trigeminal axons (Ozdinler and Erzurumlu 2002). In the developing neocortex, Slit1 is probably an endogenous dendritic branching factor for both pyramidal and nonypramidal neurons (Whitford et al. 2002). Exposure to Slit1 leads to increased dendritic growth and branching, whereas inhibition of Slit-Robo interactions attenuates dendritic branching. Thus, Slit-Robo interactions may exert a significant influence over the specification of cortical neuron morphology by regulating both axon guidance and dendritic patterning (Whitford et al. 2002).

Acknowledgements. This work was supported by grant no. 0108115 to M.J. and grant no. 0108118 to I.K. from the Croatian Ministry of Science and Technology. The authors are grateful to Marissa Kellogg for critical comments and revision of the text.

References

Adams RH, Betz H, Püschel AW (1996) A novel class of murine semaphorins with homology to thrombospondin is differentially expressed in early embryos. Mech Dev 57:33–45

Allendoerfer KL, Shatz CJ (1994) The subplate, a transient neocortical structure: its role in the development of connections between thalamus and cortex. Annu Rev Neurosci 17:185–218

Auladell C, Perez-Sust P, Super H, Soriano E (2000) The early development of thalamocortical and corticothalamic projections in the mouse. Anat Embryol (Berl) 201:169–179

Bagnard D, Lohrum M, Uziel D, Püschel AW, Bolz J (1998) Semaphorins act as attractive and repulsive guidance signals during the development of cortical projections. Development 125:5043–5053

Bagnard D, Thomasset N, Lohrum M, Püschel AW, Bolz J (2000) Spatial distributions of guidance molecules regulate chemorepulsion and chemoattraction of growth cones. J Neurosci 20:1030–1035

Bagnard D, Chounlamountri N, Püschel A, Bolz J (2001) Axonal surface molecules act in combination with semaphorin 3A during the establishment of corticothalamic projections. Cereb Cortex 11:278–285

Bagri A, Marin O, Plump AS, Mak J, Pleasure SJ, Rubenstein JLR, Tessier-Lavigne M (2002) Slit proteins prevent midline crossing and determine the dorsoventral position of major axonal pathways in the mammalian forebrain. Neuron 33:233–248

Battye R, Stevens A, Jacobs JR (1999) Axon repulsion from the midline of the *Drosophila* CNS requires slit function. Development 126:2475–2481

Behar O, Golden JA, Mashimo H, Schoen FJ, Fishman MC (1996) Semaphorin III is needed for normal patterning and growth of nerves, bones and heart. Nature 383:525–528

Bertuzzi S, Hindges R, Mui SH, O'Leary DDM, Lemke G (1999) The homeodomain protein Vax1 is required for axon guidance and major tract formation in the developing forebrain. Genes Dev 13:3092–3105

Bicknese AR, Sheppard AM, O'Leary DDM, Pearlman AL (1994) Thalamocortical axons extend along a chondroitin sulfate proteoglycan-enriched pathway coincident with the neocortical subplate and distinct from the efferent path. J Neurosci 14: 3500–3510

Bloch-Gallego E, Ezan F, Tessier-Lavigne M, Sotelo C (1999) Floor plate and netrin-1 are involved in the migration and survival of inferior olivary neurons. J Neurosci 19:4407–4420

Bolz J, Castellani V (1997) How do wiring molecules specify cortical connections? Cell Tissue Res 290:307–317

Braisted JE, Tuttle R, O'Leary DDM (1999) Thalamocortical axons are influenced by chemorepellent and chemoattractant activities localized to decision points along their path. Dev Biol 208:430–440

Braisted JE, Catalano SM, Stimac R, Kennedy TE, Tessier-Lavigne M, Shatz CJ, O'Leary DDM (2000) Netrin-1 promotes thalamic axon growth and is required for proper development of the thalamocortical projection. J Neurosci 20:5792–5801

Brose K, Bland KS, Wang KH, Arnott D, Henzel W, Goodman CS, Tessier-Lavigne M, Kidd T (1999) Slit proteins bind Robo receptors and have an evolutionarily conserved role in repulsive axon guidance. Cell 96:795–806

Castellani V, Yue Y, Gao PP, Zhou R, Bolz J (1998) Dual action of a ligand for Eph receptor tyrosine kinases on specific populations of axons during the development of cortical circuits. J Neurosci 18:4663–4672

Castellani V, Chedotal A, Schachner M, Faivre-Sarrailh C, Rougon G (2000) Analysis of the L1-deficient mouse phenotype reveals cross-talk between Sema3A and L1 signaling pathways in axonal guidance. Neuron 27:237–249

Catalano SM, Messersmith EK, Goodman CS, Shatz CJ, Chedotal A (1998) Many major CNS axon projections develop normally in the absence of Semaphorin III. Mol Cell Neurosci 11:173–182

Chédotal A, Ja DR, Ruiz M, He Z, Borrell V, De Castro F, Ezan F, Tessier-Lavigne M, Sotelo C, Soriano E (1998) Semaphorins III and IV repel hippocampal axons via two distinct receptors. Development 125:4313–4323

Chen H, Chedotal A, He Z, Goodman CS, Tessier-Lavigne M (1997) Neuropilin-2, a novel member of the neuropilin family, is a high-affinity receptor for the semaphorins Sema E and Sema IV, but not Sema III. Neuron 19:547–559

Chen H, Bagri A, Zupicich JA, Zou Y, Stoeckli E, Pleasure SJ, Lowenstein DH, Skarnes WC, Chedotal A, Tessier-Lavigne M (2000) Neuropilin-2 regulates the development of selective cranial and sensory nerves and hippocampal mossy fiber projections. Neuron 25:43–56

Chen JH, Wen L, Dupuis S, Wu JY, Rao Y (2001) The N-terminal leucine-rich regions in Slit are sufficient to repel olfactory bulb axons and subventricular zone neurons. J Neurosci 21:1548–1556

Chisholm A, Tessier-Lavigne M (1999) Conservation and divergence of axon guidance mechanisms. Curr Opin Neurobiol 9:603–615

Cloutier JF, Giger RJ, Koentges G, Dulac C, Kolodkin AL, Ginty DD (2002) Neuropilin-2 mediates axonal fasciculation, zonal segregation, but not axonal convergence, of primary accessory olfactory neurons. Neuron 33:877–892

Cohen NR, Taylor JST, Scott LB, Guillery RW, Soriano P, Furley AJW (1998) Errors in corticospinal axon guidance in mice lacking the neural cell adhesion molecule L1. Curr Biol 8:26–33

Colamarino SA, Tessier-Lavigne M (1995) The role of the floor plate in axon guidance. Annu Rev Neurosci 18:497–529

Culotti JG, Kolodkin AL (1996) Functions of netrins and semaphorins in axon guidance. Curr Opin Neurobiol 6:81–88

Dahme M, Bartsch U, Martini R, Anliker B, Schachner M, Mantei N (1997) Disruption of mouse L1 gene leads to malformations of the nervous system. Nat Genet 17:346–352

De Castro F, Hu L, Drabkin H, Sotelo C, Chedotal A (1999) Chemoattraction and chemorepulsion of olfactory bulb axons by different secreted semaphorins. J Neurosci 19:4428–4436

Demyanenko GP, Tsai AY, Maness PF (1999) Abnormalities in neuronal process extension, hippocampal development, and the ventricular system of L1 knockout mice. J Neurosci 19:4907–4920
Dodd J, Morton SB, Karagogeos D, Yamamoto M, Jessell TM (1988) Spatial regulation of axon glycoprotein expression on subsets of embryonic spinal neurons. Neuron 1:105–116
Donoghue MJ, Rakic P (1999a) Molecular gradients and compartments in the embryonic primate cerebral cortex. Cereb Cortex 9:586–600
Donoghue MJ, Rakic P (1999b) Molecular evidence for the early specification of presumptive functional domains in the embryonic primate cerebral cortex. J Neurosci 19:5967–5979
Donovan SL, Mamounas LA, Andrews AM, Blue ME, McCasland JS (2002) GAP-43 is critical for normal development of the serotonergic innervation in forebrain. J Neurosci 2289:3543–3552
Erskine L, Williams SE, Brose K, Kidd T, Rachel RA, Goodman CS, Tessier-Lavigne M, Mason CA (2000) Retinal ganglion cell axon guidance in the mouse optic chiasm: expression and function of robos and slits. J Neurosci 20:4975–4982
Fazeli A, Dickinson SL, Hermiston ML, Tighe RV, Steen RG, Small CG, Stoeckli ET, Keino-Masu K, Masu M, Rayburn H, Simons J, Bronson RT, Gordon JI, Tessier-Lavigne M, Weinberg RA (1997) Phenotype of mice lacking functional Deleted in colorectal cancer (Dcc) gene. Nature 386:796–804
Feldheim DA, Vanderhaeghen P, Hansen MJ, Frisén J, Lu Q, Barbacid M, Flanagan JG (1998) Topographic guidance labels in a sensory projection to the forebrain. Neuron 21:1303–1313
Flanagan JG, Vanderhaeghen P (1998) The Ephrins and Eph receptors in neural development. Annu Rev Neurosci 21:309–345
Fujimori KE, Takeuchi K, Yazaki T, Uyemura K, Nojyo Y, Tamamaki N (2000) Expression of L1 and TAG-1 in the corticospinal, callosal, and hippocampal commissural neurons in the developing rat telencephalon as revealed by retrograde and in situ hybridization double labeling. J Comp Neurol 417:275–288
Gao PP, Zhang JH, Yokoyama M, Racey B, Dreyfus CF, Black IB, Zhou R (1996) Regulation of topographic projection in the brain: Elf-1 in the hippocamposeptal system. Proc Natl Acad Sci USA 93:11161–11166
Gao PP, Yue Y, Zhang JH, Cerretti DP, Levitt P, Zhou R (1998) Ephrin-dependent growth and pruning of hippocampal axons. Proc Natl Acad Sci USA 95:5329–5334
Giger RJ, Wolfer DP, De Wit GMJ, Verhaagen J (1996) Anatomy of rat semaphorin III/collapsin-1 mRNA expression and relationship to developing nerve tracts during neuroembryogenesis. J Comp Neurol 375:378–392
Giger RJ, Urquhart ER, Gillespie SKH, Levengood DV, Ginty DD, Kolodkin AL (1998) Neuropilin-2 is a receptor for semaphorin IV: insight into the structural basis of receptor function and specificity. Neuron 21:1079–1092
Giger RJ, Cloutier JF, Sahay A, Prinjha RK, Levengood DV, Moore SE, Pickering S, Simmons D, Rastan S, Walsh FS et al (2000) Neuropilin-2 is required in vivo for selective axon guidance responses to secreted semaphorins. Neuron 25:29–41
Goodman CS (1996) Mechanisms and molecules that control growth cone guidance. Annu Rev Neurosci 19:341–377
Hallonet M, Hollemann T, Wehr R, Jenkins NA, Copeland NG, Pieler T, Gruss P (1998) Vax1 is a novel homeobox-containing gene expressed in the developing anterior ventral forebrain. Development 125:2599–2610

He Z, Tessier-Lavigne M (1997) Neuropilin is a receptor for the axonal chemorepellent Semaphorin III. Cell 90:739–751

Henkemeyer M, Orioli D, Henderson JT, Saxton TM, Roder J, Pawson T, Klein R (1996) Nuk controls pathfinding of commissural axons in the mammalian central nervous system. Cell 86:35–46

Holmes GP, Negus K, Burridge L, Raman S, Alger E, Yamada T, Little MH (1998) Distinct but overlapping expression patterns of two vertebrate slit homologs implies functional roles in CNS development and organogenesis. Mech Dev 79:57–72

Hong K, Nishiyama M, Henley J, Tessier-Lavigne M, Poo M (2000) Calcium signalling in the guidance of nerve growth by netrin-1. Nature 403:93–98

Hopker VH, Shewan D, Tessier-Lavigne M, Poo M, Holt C (1999) Growth-cone attraction to netrin-1 is converted to repulsion by laminin-1. Nature 401:69–73

Hu H (1999) Chemorepulsion of neuronal migration by Slit2 in the developing mammalian forebrain. Neuron 23:703–711

Hu H (2001) Cell-surface heparan sulfate is involved in the repulsive guidance activities of Slit2 protein. Nat Neurosci 4:695–701

Hu H, Rutishauser U (1996) A septum-derived chemorepulsive factor for migrating olfactory interneuron precursors. Neuron 16:933–940

Itoh A, Miyabayashi T, Ohno M, Sakano S (1998) Cloning and expressions of three mammalian homologues of Drosophila slit suggest possible roles for Slit in the formation and maintenance of the nervous system. Mol Brain Res 62:175–186

Kamiguchi H, Hlavin ML, Yamasaki M, Lemmon V (1998) Adhesion molecules and inherited diseases of the human nervous system. Annu Rev Neurosci 21:97–125

Kawakami A, Kitsukawa T, Takagi S, Fujisawa H (1996) Developmentally regulated expression of a cell surface protein, neuropilin, in mouse nervous system. J Neurobiol 29:1–17

Keino-Masu K, Masu M, Hinck L, Leonardo ED, Chan SS, Culotti JG, Tessier-Lavigne M (1996) Deleted in colorectal cancer (DCC) encodes a netrin receptor. Cell 87:175–185

Kennedy TE, Serafini T, de la Torre JR, Tessier-Lavigne M (1994) Netrins are diffusible chemotropic factors for commissural axons in the embryonic spinal cord. Cell 78:425–435

Kidd T, Brose K, Mitchell KJ, Fetter RD, Tessier-Lavigne M, Goodman CS, Tear G (1998a) Roundabout controls axon crossing of the CNS midline and defines a novel subfamily of evolutionarily conserved guidance receptors. Cell 92:205–215

Kidd T, Russell C, Goodman CS, Tear G (1998b) Dosage-sensitive and complementary functions of roundabout and commissureless control axon crossing of the CNS midline. Neuron 20:25–33

Kidd T, Bland KS, Goodman CS (1999) Slit is the midline repellent for the robo receptor in *Drosophila*. Cell 96:785–794

Kitsukawa T, Shimizu M, Sanbo M, Hirata T, Taniguchi M, Bekku Y, Yagi T, Fujisawa H (1997) Neuropilin-semaphorin III/D-mediated chemorepulsive signals play a crucial role in peripheral nerve projection in mice. Neuron 19:995–1005

Klein R (2001) Excitatory Eph receptors and adhesive ephrin ligands. Curr Opin Cell Biol 13:196–203

Knöll B, Drescher U (2002) Ephrin-As as receptors in topographic projections. Trends Neurosci 25:145–149

Kobayashi H, Koppel AM, Luo Y, Raper JA (1997) A role for collapsin-1 in olfactory and cranial sensory axon guidance. J Neurosci 17:8339–8352

Kolodkin AL, Levengood DV, Rowe EG, Tai UT, Giger RJ, Ginty DD (1997) Neuropilin is a semaphorin III receptor. Cell 90:753–762

Kostović I, Judaš M (1995) Prenatal development of the cerebral cortex. In: Chervenak FA, Kurjak A, Comstock CH (eds) Ultrasound and the fetal brain. Parthenon Publishing Group, New York, pp 1–26
Kostović I, Judaš M (2002) Correlation between the sequential ingrowth of afferents and transient patterns of cortical lamination in preterm infants. Anat Rec 267:1–6
Kostović I, Rakic P (1990) Developmental history of the transient subplate zone in the visual and somatosensory cortex of the macaque monkey and human brain. J Comp Neurol 297:441–470
Kostović I, Judaš M, Petanjek Z, Šimić G (1995) Ontogenesis of goal-directed behavior: anatomo-functional considerations. Int J Psychophysiol 19:85–102
Kostović I, Judaš M, Radoš M, Hrabač P (2002) Laminar organization of the human fetal cerebrum revealed by histochemical markers and magnetic resonance imaging. Cereb Cortex 12:536–544
Kullander K, Mather NK, Diella F, Dottori M, Boyd AW, Klein R (2001) Kinase-dependent and kinase-independent functions of EphA4 receptors in major axons tract formation in vivo. Neuron 29:73–84
Lauderdale JD, Pasquali SK, Fazel R, van Eeden FJ, Schauerte HE, Haffter P, Kuwada JY (1998) Regulation of netrin-1a expression by hedgehog proteins. Mol Cell Neurosci 11:194–205
Leighton PA, Mitchell KJ, Goodrich LV, Lu X, Pinson K, Scherz P, Skarnes WC, Tessier-Lavigne M (2001) Defining brain wiring patterns and mechanisms through gene trapping in mice. Nature 410:174–179
Letourneau PC, Condic ML, Snow DM (1994) Interactions of developing neurons with the extracellular matrix. J Neurosci 14:915–928
Li HS, Chen JH, Wu W, Fagaly T, Zhou L, Yuan W, Dupuis S, Jiang ZH, Nash W, Gick C et al (1999) Vertebrate slit, a secreted ligand for the transmembrane protein roundabout, is a repellent for olfactory bulb axons. Cell 96:807–818
Liu BP, Strittmatter SM (2001) Semaphorin-mediated axonal guidance via Rho-related G proteins. Curr Opin Cell Biol 13:619–626
Livesey FJ, Hunt SP (1997) Netrin and netrin receptor expression in the embryonic mammalian nervous system suggests roles in retina, striatal, nigral, and cerebellar development. Mol Cell Neurosci 8:417–429
Logan C, Wizenmann A, Drescher U, Monschau B, Bonhoeffer F, Lumsden A (1996) Rostral optic tectum acquires caudal characteristics following ectopic engrailed expression. Curr Biol 6:1006–1014
Mackarehtschian K, Lau CK, Caras I, McConnell SK (1999) Regional differences in the developing cerebral cortex revealed by Ephrin-A5 expression. Cereb Cortex 9:601–610
Mann F, Zhukareva V, Pimenta A, Levitt P, Bolz J (1998) Membrane-associated molecules guide limbic and nonlimbic thalamocortical projections. J Neurosci 18: 9409–9419
Marillat V, Cases O, Nguyen Ba-Charvet KT, Tessier-Lavigne M, Sotelo C, Chédotal A (2002) Spatiotemporal expression patterns of Slit and Robo genes in the rat brain. J Comp Neurol 442:130–155
Mark MD, Lohrum M, Püschel AW (1997) Patterning neuronal connections by chemorepulsion: the semaphorins. Cell Tissue Res 290:299–306
McConnell SK, Ghosh A, Shatz CJ (1989) Subplate neurons pioneer the first axon pathway from the cerebral cortex. Science 245:978–982
Mellitzer G, Xu Q, Wilkinson DG (2000) Control of cell behavior by signaling through Eph receptors and ephrins. Curr Opin Neurobiol 10:400–408

Métin C, Godement P (1996) The ganglionic eminence may be an intermediate target for corticofugal and thalamocortical axons. J Neurosci 16:3219–3235
Métin C, Deleglise D, Serafini T, Kennedy TE, Tessier-Lavigne M (1997) A role for netrin-1 in the guidance of cortical efferents. Development 124:5063–5074
Miller B, Sheppard AM, Bicknese AR, Pearlman AL (1995) Chondroitin sulfate proteoglycans in the developing cerebral cortex: the distribution of neurocan distinguishes forming afferent and efferent axonal pathways. J Comp Neurol 355:615–628
Ming GL, Song HJ, Berninger B, Holt CE, Tessier-Lavigne M, Poo MM (1997) cAMP-dependent growth cone guidance by netrin-1. Neuron 19:1225–1235
Molnar Z, Blakemore C (1995) How do thalamic axons find their way to the cortex? Trends Neurosci 18:389–397
Molnar Z, Adams R, Blakemore C (1998a) Mechanisms underlying the early establishment of thalamocortical connections in the rat. J Neurosci 18:5723–5745
Molnar Z, Adams R, Goffinet AM, Blakemore C (1998b) The role of the first postmitotic cortical cells in the development of thalamocortical innervation in the reeler mouse. J Neurosci 18:5746–5765
Mueller BK (1999) Growth cone guidance: first steps towards a deeper understanding. Annu Rev Neurosci 22:351–388
Nakamura F, Kalb RG, Strittmatter SM (2000) Molecular basis of semaphorin-mediated axon guidance. J Neurobiol 44:219–229
Nguyen Ba-Charvet KT, Brose K, Marillat V, Kidd T, Goodman CS, Tessier-Lavigne M, Sotelo C, Chedotal A (1999) Slit2-mediated chemorepulsion and collapse of developing forebrain axons. Neuron 22:463–473
Nguyen-Ba-Charvet KT, Plump AS, Tessier-Lavigne M, Chédotal A (2002) Slit1 and Slit2 proteins control the development of the lateral olfactory tract. J Neurosci 22(13):5473–5480
O'Leary DDM, Wilkinson DG (1999) Eph receptors and ephrins in neural development. Curr Opin Neurobiol 9:65–73
Orioli D, Henkemeyer M, Lemke G, Klein R, Pawson T (1996) Sek4 and Nuk receptors cooperate in guidance of commissural axons and in palate formation. EMBO J 15:6035–6049
Ozdinler PH, Erzurumlu RS (2002) Slit2, a branching-arborization factor for sensory axons in the mammalian CNS. J Neurosci 22:4540–4549
Pearlman AL, Sheppard AM (1996) Extracellular matrix in early cortical development. Prog Brain Res 108:119–134
Pini A (1993) Chemorepulsion of axons in the developing mammalian central nervous system. Science 261:95–98
Plump AS, Erskine L, Sabatier C, Brose K, Epstein CJ, Goodman CS, Mason CA, Tessier-Lavigne M (2002) Slit1 and Slit2 cooperate to prevent premature midline crossing of retinal axons in the mouse visual system. Neuron 33:219–232
Polleux F, Giger RJ, Ginty DD, Kolodkin AL, Ghosh A (1998) Patterning of cortical efferent projections by semaphorin-neuropilin interactions. Science 282:1904–1906
Polleux F, Morrow T, Ghosh A (2000) Semaphorin 3A is a chemoattractant for cortical apical dendrites. Nature 404:567–573
Rabacchi SA, Solowska JM, Kruk B, Luo Y, Raper JA, Baird DH (1999) Collapsin-1/semaphorin-III/D is regulated developmentally in Purkinje cells and collapses pontocerebellar mossy fiber neuronal growth cones. J Neurosci 19:4437–4448
Rakic P (1982) Early developmental events: cell lineages, acquisition of neuronal positions, and areal and laminar development. Neurosci Res Prog Bull 20:439–451

Raper JA (2000) Semaphorins and their receptors in vertebrates and invertebrates. Curr Opin Neurobiol 10:88–94

Richards LJ, Koester SE, Tuttle R, O'Leary DDM (1997) Directed growth of early cortical axons is influenced by a chemoattractant released from an intermediate target. J Neurosci 17:2445–2458

Ringstedt T, Braisted JE, Brose K, Kidd T, Goodman C, Tessier-Lavigne M, O'Leary DDM (2000) Slit inhibition of retinal axon growth and its role in retinal axon pathfinding and innervation patterns in the diencephalon. J Neurosci 20:4983–4991

Schwarting GA, Kostek C, Ahmad N, Dibble C, Pays L, Püschel AW (2000) Semaphorin 3A is required for guidance of olfactory axons in mice. J Neurosci 20:7691–7697

Semaphorin Nomenclature Committee (1999) Unified nomenclature for the semaphorins/collapsins. Cell 97:551–552

Serafini T, Kennedy TE, Galko MJ, Mirzayan C, Jessell TM, Tessier-Lavigne M (1994) The netrins define a family of axon outgrowth-promoting proteins homologous to *C. elegans* UNC-6. Cell 78:409–424

Serafini T, Colamarino SA, Leonardo ED, Wang H, Beddington R, Skarnes WC, Tessier-Lavigne M (1996) Netrin-1 is required for commissural axon guidance in the developing vertebrate nervous system. Cell 87:1001–1014

Šestan N, Rakic P, Donoghue MJ (2000) Independent parcellation of the embryonic visual cortex and thalamus revealed by combinatorial Eph/ephrin gene expression. Curr Biol 11:39–43

Shen Y, Mani S, Donovan SL, Schwob JE, Meiri KF (2002) Growth-associated protein-43 is required for commissural axon guidance in the developing vertebrate nervous system. J Neurosci 22:239–247

Sheppard AM, Hamilton SK, Pearlman AL (1991) Changes in the distribution of extracellular matrix components accompany early morphogenetic events of mammalian cortical development. J Neurosci 11:3928–3942

Shigetani Y, Funahashi JI, Nakamura H (1997) En-2 regulates the expression of the ligands for Eph type tyrosine kinases in chick embryonic tectum. Neurosci Res 27:211–217

Shirasaki R, Tamada A, Katsumata R, Murakami F (1995) Guidance of cerebellofugal axons in the rat embryo: directed growth toward the floor plate and subsequent elongation along the longitudinal axis. Neuron 14:961–972

Shu T, Richards LJ (2001) Cortical axon guidance by the glial wedge during the development of the corpus callosum. J Neurosci 21:2749–2758

Silver J, Ogawa MY (1983) Postnatally induced formation of the corpus callosum in acallosal mice on glia-coated cellulose bridges. Science 220:1067–1069

Silver J, Edwards MA, Levitt PJ (1993) Immunocytochemical demonstration of early appearing astroglial structures that form boundaries and pathways along axon tracts in the fetal brain. J Comp Neurol 328:415–436

Skaliora I, Singer W, Betz H, Püschel AW (1998) Differential patterns of semaphorin expression in the developing rat brain. Eur J Neurosci 10:1215–1229

Skaliora I, Adams R, Blakemore C (2000) Morphology and growth patterns of developing thalamocortical axons. J Neurosci 20:3650–3662

Song HJ, Poo MM (2001) The cell biology of neuronal navigation. Nat Cell Biol 3:EB1–EB8

Steup A, Ninnemann O, Savaskan NE, Nitsch R, Püschel AW, Skutella T (1999) Semaphorin D acts as a repulsive factor for entorhinal and hippocampal neurons. Eur J Neurosci 11:729–734

Tamagnone L, Comoglio PM (2000) Signaling by semaphorin receptors: cell guidance and beyond. Trends Cell Biol 10:377–383

Taniguchi M, Yuasa S, Fujisawa H, Naruse I, Saga S, Mishina M, Yagi T (1997) Disruption of semaphorin III/D gene causes severe abnormality in peripheral nerve projection. Neuron 19:519–530

Tessier-Lavigne M, Goodman CS (1996) The molecular biology of axon guidance. Science 274:1123–1133

Tuttle R, Nakagawa Y, Johnson JE, O'Leary DDM (1999) Defects in thalamocortical axon pathfinding correlate with altered cell domains in Mash-1-deficient mice. Development 126:1903–1916

Varela-Echavarria A, Tucker A, Püschel AW, Guthrie S (1997) Motor axon subpopulations respond differentially to the chemorepellents netrin-1 and semaphorin D. Neuron 18:193–207

Wang KH, Brose K, Arnott D, Kidd T, Goodman CS, Henzel W, Tessier-Lavigne M (1999) Biochemical purification of a mammalian slit protein as a positive regulator of sensory axon elongation and branching. Cell 96:771–784

Whitford KL, Marillat V, Stein E, Goodman CS, Tessier-Lavigne M, Chédotal A, Ghosh A (2002) Regulation of cortical dendrite development by Slit-Robo interactions. Neuron 33:47–61

Wolfer DP, Henehan-Beatty A, Stoeckli ET, Sonderegger P, Lipp HP (1994) Distribution of TAG-1/axonin-1 in fibre tracts and migratory streams of the developing mouse nervous system. J Comp Neurol 345:1–32

Wu W, Wong K, Chen J, Jiang Z, Dupuis S, Wu JY, Rao Y (1999) Directional guidance of neuronal migration in the olfactory system by the protein Slit. Nature 400:331–336

Yee KT, Simon HH, Tessier-Lavigne M, O'Leary DDM (1999) Extension of long leading processes and neuronal migration in the mammalian brain directed by the chemoattractant netrin-1. Neuron 24:607–622

Yokoyama N, Romero MI, Cowan CA, Galvan P, Helmbacher F, Charnay P, Parada LF, Henkemeyer M (2001) Forward signaling mediated by ephrin-B3 prevents contralateral corticospinal axons from recrossing the spinal cord midline. Neuron 29:85–97

Yuan W, Zhou L, Chen JH, Wu JY, Rao Y, Ornitz DM (1999) The mouse SLIT family: secreted ligands for ROBO expressed in patterns that suggest a role in morphogenesis and axon guidance. Dev Biol 212:290–306

Zhang JH, Cerretti DP, Yu T, Flanagan JG, Zhou R (1996) Detection of ligands in regions anatomically connected to neurons expressing the Eph receptor Bsk: potential roles in neuron-target interaction. J Neurosci 16:7182–7192

Zhou L, White A, Lentz SI, Wright DE, Fisher DA, Snider WD (1997) Cloning and expression of a novel murine semaphorin with structure similarity to insect semaphorin 1. Mol Cell Neurosci 9:26–41

Zhou C, Qiu Y, Pereira FA, Crair MC, Tsai SY, Tsai MJ (1999) The nuclear orphan receptor COUP-TFI is required for differentiation of subplate neurons and guidance of thalamocortical axons. Neuron 24:847–859

Zhu Y, Li HS, Zhou L, Wu JY, Rao Y (1999) Cellular and molecular guidance of GABA-ergic neuronal migration from the striatum to the neocortex. Neuron 23:473–485

Zou Y, Stoeckli E, Chen H, Tessier-Lavigne M (2000) Squeezing axons out of the gray matter: a role for slit and semaphorin proteins from midline and ventral spinal cord. Cell 102:363–375

Zou Z, Horowitz LF, Montmayeur JP, Snapper S, Buck LB (2001) Genetic tracing reveals a stereotyped sensory map in the olfactory cortex. Nature 414:173–179

Phosphorylation Pattern of tau Associated with Distinct Changes of the Growth Cone Cytoskeleton

G. Šimić,[1] A. Diana,[2] P. R. Hof[3]

1 Historical Note

That both axons and dendrites grow at specialized terminations called growth cones was recognized already by Santiago Ramon y Cajal (Ramon y Cajal 1890). After inventing the technique of tissue culture, Ross G. Harrison confirmed Cajal's inference that growth cones represent the elongating tips of axons and first described a living growth cone that is moving by local extension and retraction (Harrison 1907).

2 Neuronal Cytoskeleton and Microtubule-Associated Proteins

In order to establish a correct arrangement of connections, developing nerve cells evolved to have the unique ability to place themselves in strategic positions by extending their processes. In neurons projecting elongated axons over enormous distances, the cytoskeletal system in the cytoplasm acts as a dynamic scaffold. It consists of the three major types of filamentous organelles: large (microtubules, MTs), intermediate (neurofilaments) and small filaments (microfilaments).

MTs are composed of subunits that are heterodimers of alpha- and beta-tubulin monomers. Lateral interactions hold alpha- and beta-tubulin in a stable complex, while longitudinal contacts between subunits link them head to tail into a straight column called a protofilament. Protofilaments associate side by side into an MT cylinder

[1] Department of Neuroscience, Croatian Institute for Brain Research, Zagreb University Medical School, Šalata 12, Zagreb 10000, Croatia
[2] Department of Cytomorphology, University of Cagliari, 09042 Monserrato, Cagliari, Italy
[3] Neurobiology of Aging Laboratories and Fishberg Research Center for Neurobiology, Department of Geriatrics and Adult Development, and Department of Ophthalmology, Mount Sinai School of Medicine, New York, New York 10029, USA

Progress in Molecular and Subcellular Biology, Vol. 32
Ivica Kostović (Ed.)

measuring 24 nm in diameter. Because of their tube-like structure that varies in length from a fraction of a micrometer to hundreds of micrometers, MTs are much stiffer than either neurofilaments or microfilaments. MTs exhibit a polarity in their structure which arises from the head-to-tail arrangement of alpha- and beta-tubulin in a protofilament.

In order to grow and maintain their long axons, neurons must have an internal core of stable MTs. The preferred assembly end of MTs is designated the plus end, while the end that assembles more slowly is termed the minus end. Both assembly and disassembly of MTs occur preferentially at the plus, distal, end of an axon (Lodish et al. 1997). Most, if not all, axonal MTs are initially assembled in the neuron soma, and subsequently are actively transported into the axon by specific motor proteins (Slaughter et al. 1997). This transport provides a steady supply of new MTs for the growing axon and contributes to the plus-end distal polarity orientation of axonal MTs by conveying MTs specifically with their plus end leads (Baas and Brown 1997). Regional differences in MTs dynamic in the axon stem from the observation that an enzyme that removes tyrosine, a detyrosinase, acts on alpha-tubulin that is incorporated into MTs. Because detyrosination is a polymer-specific modification of alpha-tubulin that accumulates with time, the levels of tyrosinated alpha-tubulin constitute a reflection of the age, and hence, dynamic properties of the polymer. Using this approach, it has been shown that individual MTs are staggered along the length of the axon and have a labile domain situated at the plus end of a stable domain (Baas and Black 1990). As a result, the most distal region of the axon that is nearest to the advancing growth cone consists entirely of labile domains, while the main shaft consists of a mixture of labile and stable domains (Ahmad et al. 1993).

MTs tend to form bundles in which various kinds of cross-bridge structures have been observed (Hirokawa et al. 1985). Microtubule-associated proteins (MAPs), a group of filamentous proteins, represent major components of these extensive cross-bridge structures associated with MTs. Based on sequence analysis, MAPs are grouped into two types: type I large MAPs (MAP1A and MAP1B) that can be found in both dendrites and axons, and contain several repeats of the amino acid sequence KKEX, which is implicated as a binding site for negatively charged tubulin. This sequence is postulated to neutralize the charge repulsion between tubulin subunits and stabilize the polymer. Type II MAPs include MAP2 (high-molecular MAP2A and MAP2B, and low-molecular MAP2C and MAP2D), nonneuronal MAP4, and tau. All of these proteins are characterized by three or four repeats (R1-R4) of an

18-residue sequence in the MT-binding domain. The principal members of neuronal MAPs in the developing brain are tau proteins (Cleveland et al. 1977), MAP1B and, to a lesser degree, MAP2.

tau is a component of short, 18-nm-long cross-bridge "arms" between MTs in axons, whereas MAP1B is a constituent of longer ones (Sato-Yoshitake et al. 1989). MAP2 cross-link MTs in dendrites. When MAPs coat the outer wall of an MT, tubulin subunits are unable to dissociate.

3
Neuronal Growth Cone Composition and Translocation

The neuronal growth cone is the tip of an elongated axon or dendrite that is intimately associated with pathway formation and synaptogenesis. At the light-microscopic level, growth cones look like flattened clubs or "fingers", called lamellipodia, with numerous fine membrane extensions known as microspikes or filopodia. The lamellipodia extend forward while remaining attached to the substrate. Filopodia extend and retract rapidly, often moving from side to side with no attachment to the substrate, as if probing the environment.

At the ultrastructural level, the central region of the growth cone is comprised of MTs and neurofilaments surrounded by microfilaments, smooth endoplasmic reticulum, mitochondria, vesicles, and a few lysosomes (Yamada et al. 1971). Towards the distal border of the growth cone these organelles become less frequent, so that beneath the plasma membrane there is a relatively organelle-free zone in which linear polymers of actin microfilaments predominate (Gordon-Weeks 1987).

Growth cone translocation is a key event in axonal elongation. The sequence of cytoskeleton-dependent advance of the growth cone can be subdivided into the four major stages (Tanaka and Sabry 1995). In stage 1 (exploration) the filopodia and lamellipodia are highly dynamic, and their shapes are determined mainly by the organization of actin microfilaments within them, while MTs are localized in the central domain of the growth cone. In stage 2 (invasion) unstable MTs enter into the periphery of the growth cone. Stage 3 (stabilization) is characterized by organizing MTs to form a bundle within the growth cone. This is the crucial step during which tau and MAP1B proteins are considered to zip MTs from the proximal side of the neck of the advancing growth cone, and thus convert the dynamically unstable MTs into highly organized parallel arrays of axonal MTs (Takei et al. 2000). Then, in stage 4 (axon formation), the plasma membrane of the growth cone collapses around the MTs bundle to generate a new axonal tube.

4 tau Gene Organization and Regulation

The human tau is encoded by a single gene located over 100 kb on the long arm of chromosome 17 at position 17q21 and is expressed primarily in neurons (Binder et al. 1985; Neve et al. 1986). The tau gene is regulated by a single neuron-specific promoter that has three binding sites for transcriptional factors. Mutations within any of these three binding sites decreases transcriptional activity of the tau gene (Heicklen-Klein and Ginzburg 2000). Tau promoter activity increases significantly with differentiation, specifically during neurite initiation and outgrowth in cultured neurons and neural cell lines (Drubin et al. 1985; Ferreira et al. 1989; Brion et al. 1994; Smith et al. 1995; Heicklen-Klein et al. 2000), which suggests an important role of tau in axon formation.

The human tau primary transcript contains 13 exons of which exons –1 and 14 are not translated. Exons 1, 4, 5, 7, 9, 11, 12 and 13 are constitutive, while exons 2, 3 and 10 alternatively splice and give rise to a family of six isoforms which range from 352 to 441 amino acids (Buée et al. 2000). Since they are differentially expressed during development, each of these isoforms is likely to have particular regulation and physiological role. In fetal human brain, there is only one isoform referred to as fetal tau, which is absent of regions encoded by the exons 2, 3 and 10 (Andreadis et al. 1992) and, therefore the shortest 352-amino acids isoform. The adult tau isoforms appear postnatally (Goedert et al. 1989; Kosik et al. 1989) and are differentially distributed in neuronal subpopulations of human brain. For example, tau mRNAs containing exon 10 are not found in granular cells of the dentate gyrus (Goedert et al. 1989). It has been suggested that the unique expression pattern of tau isoforms in the human central nervous system implicates a possible link to the particular vulnerability of humans to neurodegenerative disorders with tauopathy, especially Alzheimer's disease (Janke et al. 1999).

5 Structure of tau Proteins

The tau protein is a heat-stable phosphoprotein that looks like a random coil. It has about 17% of serine and threonine residues and can be divided in two functional domains: the basic MT binding domain (towards the C-terminus) and the acidic projection domain (towards the N-terminus) (Buée et al. 2000).

The microtubule binding domain regulates the rate of MT polymerization. Tau proteins bind MTs through highly conserved repetitive domains R1–R4 that are encoded by exons 9–12 (Lee et al. 1989). Adult tau isoforms with 4R (R1–R4) are about 40-fold more efficient at promoting MT assembly than the fetal isoform that is lacking exon 10 and thus having only 3R (R1, R3 and R4; Goedert and Jakes 1990; Butner and Kirschner 1991; Panda et al. 1995). It is thought that the absence of expression of the R1–R2 inter-repeat region during fetal development allows for the cytoskeletal plasticity required of growing immature neurons (Goode and Feinstein 1994). Through their MT binding domain, tau proteins also bind to RNA.

The projection domain is so called because in the electron microscope it appears as a filamentous 'arm' that projects from the wall of the microtubule. The projection domain of tau can bind to neurofilaments, neural plasma membrane or other MTs. The variety of potential binding partners suggests that function of tau is to integrate MTs with other cytoplasmic structures. Interaction with neurofilaments interconnect MTs with other cytoskeletal components and restricts the flexibility of MTs (Miyata et al. 1986). When binding to other MTs, projection domain length controls how far apart MTs are spaced (Chen et al. 1992). Through binding of its proline-rich sequence with the SH3 domains of src kinases, PLC-gamma, spectrin and actin (Selden and Pollard 1983; Brandt et al. 1995), the tau projection domain interacts with neural plasma membrane.

6
Post-Translational Modifications of tau Proteins

The most common post-translational modifications of the tau proteins are phosphorylation and O-glycosylation. Phosphorylation changes the shape of the tau molecule (Hagestedt et al. 1989) and regulates its biological activity. In turn, tau proteins change physical characteristics of MTs such as rigidity, length, stability and integrative capacity with other organelles. Tau in a more extensively phosphorylated state has a decreased affinity for MTs and is less able to promote MT assembly compared to dephosphorylated tau (Lindwall and Cole 1984). Out of 80 putative Ser or Thr phosphorylation sites on the longest brain tau isoform, so far at least 30 of them have been characterized by monoclonal antibodies, mass spectrometry and sequencing (Buée et al. 2000). Most of the phosphorylation sites are on Ser-Pro and Thr-Pro motives, but a number of sites on other residues have also been reported (Morishima-Kawashima et al. 1995). A different state of tau phosphorylation results from the activity of probably more than 15 different specific kinases (such as proline-directed glycogen synthase kinase 3, mitogen- and stress-activated kinases, cyclin-dependent kinases, and nonproline directed microtubule affinity-regulating kinase, calcium/calmodulin kinase II, cyclic-AMP-dependent kinase A, casein kinase II and phosphorylase K), and at least four different phosphatases (Ser/Thr phosphatase 1, 2A, 2B and 2C; Buée and Delacourte 2001).

Phosphorylation of tau proteins is developmentally regulated (Rösner et al. 1995; Riederer et al. 2001). It is high in the fetal period and decreases with advancing age due to phosphatase activation (Mawal-Dewan et al. 1994; Dudek and Johnson 1995; Rösner et al. 1995). However, after about 30 years of age, the plasticity burden in limbic and basal forebrain areas, together with additional genetic (mutation of amyloid precursor protein, presenilins and cytochrome-oxidases, polymorphism of apolipoprotein E-gene) and environmental (head trauma) factors that interfere with synaptic plasticity, cause the phosphorylation to increase again (perhaps this time together with increased cleavage of tau; Mesulam 1999). It is currently thought that the fibrillogenesis of diffuse beta-amyloid deposits is the main mechanism that induces physical damage to axons, thus compromising axoplasmatic flow (Haas and Steiner 2001). Neuronal response to this kind of injury results in axonal sprouting, tau mobilization and excessive phosphorylation (Vickers et al. 2000). Hyperphosphorylation promotes polymerization of unbound tau and formation of straight and paired helical filaments that form neurofibrillary tangles, dystrophic neurites and neuropil threads (Grundke-Iqbal et al. 1986; Ihara et al. 1986; Kosik et al. 1986; Wood et al. 1986; Buée et al. 2000). These pathological changes characterize Alzheimer's disease and, to a lesser degree, a number of other neurodegenerative diseases and normal aging (Šimić et al. 1998). It seems that it takes about 20–30 years from the earliest neuronal changes caused by light hyperphosphorylation to the development of obvious symptoms of sporadic Alzheimer's disease, because such neurons are only visible in adults over 30 (Braak et al. 1994). Although it has become evident that tau is highly phosphorylated in the developing brain in a manner similar to the phosphorylation state in Alzheimer's disease (Kanemaru et al. 1992; Brion et al. 1993; Bramblett et al. 1993; Goedert et al. 1993; Kenessey and Yen 1993; Watanabe et al. 1993; Brion et al. 1995), the precise roles of microtubule-associated fetal kinases and their respective phosphorylation epitopes in this process have yet to be elucidated (Jenkins and Johnson 1997; Arendt 2000; Mehler and Gokhan 2001; Riederer et al. 2001).

O-glycosylation is, in a way, reciprocal to phosphorylation of tau (Haltiwanger et al. 1997). It is characterized by the addition of an O-linked N-acetylglucosamine (O-GlcNAc) on Ser or Thr residues in the proximity of Pro residues. O-glycosylation of tau suggests a role of O-GlcNAc in mediating its interactions with tubulin. Presumably, glycosylation may also have a role in subcellular localization and degradation of tau proteins (Arnold et al. 1996), and maintenance of paired helical filaments (Wang et al. 1996; Goedert et al. 1996).

7
Subcellular Localization of tau

Initial studies on subcellular localization of MAPs have revealed that tau proteins are confined preferentially within the axonal, whereas high molecular weight MAPs within the somatodendritic domain (Matus et al. 1981; Binder et al. 1985). However, later studies that labeled cells with phosphorylation-independent antibodies have demonstrated that tau proteins can be found in all cell compartments, even in the nucleus (Brady et al. 1995; Uberti et al. 1997), but in different phosphorylation states. For example, in cultured chick spinal cord neurons and rat cerebral cortex neurons, the tau phosphorylated at Ser202 is concentrated in the proximal parts of axons, close to the cell body, while tau that lacks phosphorylation at this residue is concentrated in the distal axon and growth cone (Rebhan et al. 1995). Similarly, in developing cultured hippocampal neurons it has been shown that the change of phosphorylation state from soma to axon takes the form of a smooth proximo-distal gradient, with tau in the soma, immature dendrites and proximal axon being about 80% phosphorylated at the tau-1 site, and in the axonal growth cone only 20% (Mandell and Banker 1996). These studies confirmed initial suggestions that an independent regulation of multiple phosphorylation sites takes place within different subcellular domains of developing neurons (Szendrei et al. 1993; Goedert et al. 1995). This suggests that both the overall amount of phosphorylation and the precise amino acid residues bearing phosphates are important for the regulation of the function of tau proteins.

Using GFP-tagged tau constructs combined with in situ hybridization analysis, a colocalization of the targeted tau mRNA and its translated protein in the axon and growth cone has recently been demonstrated (Aronov et al. 2001). This finding indicates that tau synthesis may be increased locally in response to a local demand. Beyond contributions from mRNA localization and selective axonal transport, compartmentalization of tau has been proposed to occur also through local differential turnover (Black et al. 1996; Hirokawa et al. 1996).

8 The Role of tau Phosphorylation in Establishment of Axonal Polarity

All moving cells display polarity. Probably the earliest step in acquisition of neuronal polarity is axonogenesis. During migration of cerebellar granular cells, the trailing process elongates to form the parallel fibers, and in the cerebral cortex, the trailing process appears to become the axon. However, the axon of spinal cord motor neurons and retinal ganglion cells arises out of what had been the leading edge of the migrating neuron, suggesting that there is no simple relationship between cell migration and the establishment of neuronal polarity (Devoto 1990).

Neurons in dissociated culture initially extend exploratory neurites and, after about a day of outgrowth, become polarized. Polarity becomes evident when a single neurite exceeds the others in length. Such a stable neurite ceases to undergo the retractions and extensions characteristic of initial neurites and acquires many features of a developed axon. However, if it is cut shorter, then the next longest becomes the axon instead (Dotti and Banker 1987). A few days later, the other processes begin to elongate rapidly, and differentiate into dendrites (Dotti et al. 1988).

The subcellular mechanisms underlying these events are largely unknown, but it is believed that MAPs, and in particular tau, play a key role in this process (Kempf et al. 1996; Bassell and Singer 1997). However, although tau and other MAPs are polarized in mature neurons in vivo, they are unpolarized during early stages of axonogenesis. The simplest explanation would be the existence of a parallel axonal concentration gradient of a relevant kinase or phosphatase, but no such evidence has yet been obtained. To explain this discrepancy, a mechanism of locally controlled phosphorylation has been suggested. Namely, even if the tau kinases and phosphatases are uniformly distributed during neuronal polarization, they could be locally regulated by polarized second messengers or regulatory proteins necessary for their activation. Indeed, calcium and cAMP have already been demonstrated to spontaneously form concentration gradients in growing neurites (Kater et al. 1988; Bacskai et al. 1993), making it logical to assume that the generation and modulation of specific phosphorylation gradients and polarization of neurons may be driven by the extracellular cues present in the neuronal microenvironment (such as extracellular matrix molecules, cell surface clues, neurotransmitters, diffusible growth factors; Snow et al. 2001). Many of these cues act through tyrosine kinase- or phosphatase-linked transmembrane receptors. For example, endocytosis of the cell adhesion molecule L1 that is controlled by a phosphorylation–dephosphorylation cycle stimulated by outside-in signaling by L1 (Schaefer et al. 2002) leads to dephosphorylation of tubulin tyrosine residues (Atashi et al. 1992). Moreover, since a growth cone that migrates on an L1 substrate exhibits a steep gradient of L1-mediated adhesion (strong adhesion at the growth cone's leading edge and weak adhesion at the central domain), it can be supposed that the growth cone actively translocates cell adhesion molecules to create spatial asymmetry (Kamiguchi and Yoshihara, 2001). In favor of this hypothesis also goes the fact that similar phosphorylation gradients can be demonstrated for many other neuronal proteins, such as neurofilaments (Szaro et al. 1989).

It can be concluded that the establishment of neuronal polarity could develop after an initial random outgrowth. If a process growing in the direction of the future axon were able to elongate more rapidly than processes growing in the other directions, and the cell limited the number of axons to one, this would be sufficient to establish polarity.

9
The Role of tau Phosphorylation in Axonal Outgrowth and Morphology

While the movement of the cell body determines the position that the neuron will assume, the movements of the growth cones determine the initial morphology of the developing cell. Movements of the growth cones also determine which cells the neuron will have the opportunity to form synapses with. A link between tau phosphorylation and neurite development has been established using in vitro models, where the phosphorylation state of tau was experimentally manipulated with inhibitors or activators of various protein kinases and phosphatases. The role of tau in promoting axon growth has been also tested using transfection of cells with tau DNA, adding antisense oligonucleotides to the culture media and analyses of tau knock-out mice. Analyses of fibroblasts transfected with a tau gene have revealed that tau induces axon-like processes which are filled with bundles of MTs (Kanai et al. 1989). Conversely, with tau protein synthesis blocked by antisense oligonucleotides neurons failed to grow an axon, although the extension of initial exploratory neurites occurred normally (a similar effect was seen for dendrites when the levels of MAP2 were reduced), suggesting that development of neuronal polarity was inhibited (Caceres and Kosik 1990; Caceres et al. 1991).

Analysis of tau knock-out neurons revealed a decreased axonal diameter and decrease in the number of MTs, while neurons have extended axons indistinguishable from wild-type controls (Harada et al. 1994), possibly due to an increase in other MAPs that partially compensate for the functions of tau (DiTella et al. 1996; Tint et al. 1998). By using chromophore-assisted laser inactivation of tau, which generates an acute loss of tau function not subjected to compensation by other MAPs, inactivation of tau within growth cones caused a decrease in neurite extension rate by approximately two-fold (Liu et al. 1999). Surprisingly, it also decreased the lamellipodial size within the inactivated region by about 20%, suggesting that tau is required not only for neurite outgrowth, but also for motile behavior and elaboration at the growth cone leading edge of migrating neurons (Liu et al. 1999). Subsequent studies in tau mutant growth cones have confirmed altered localization of actin polymers (Takei et al. 2000). It has also been shown that depolymerization of filamentous actin is a prerequisite for dephosphorylated tau to bind all the way to the most distal tips of MTs in the axon (Zmuda and Rivas 2000).

By interaction with actin, tau guides site-directed MT redistribution towards the target, thus modifying growth cone morphology (Lin and Forscher 1993; Lee et al. 1998). This is a necessary event in growth cone turning (Williamson et al. 1996). Selective changes in the phosphorylation pattern of tau after actin-tau interactions have also been reported in association with neurite branching (Audesirk et al. 1997) and MT assembly in thrombin- and semaphorin-induced responses (Fritsche et al. 1999). These data provide further evidence that changes in the phosphorylation pattern of tau serve as a mechanism whereby transduction pathways of different extracellular and intracellular signals lead to distinct changes of the growth cone cytoskeleton.

References

Ahmad FJ, Pienkowski TP, Baas PW (1993) Regional differences in microtubule dynamics in the axon. J Neurosci 13:856–866

Andreadis A, Brown BK, Kosik KS (1992) Structure and novel exons of the human tau gene. Biochemistry 31:10626–10633

Arendt T (2000) Alzheimer's disease as a loss of differentiation control in a subset of neurons that retain immature features in the adult brain. Neurobiol Aging 21:783–796

Arnold CS, Johnson GV, Cole RN, Dong DLY, Lee M, Hart GW (1996) The microtubule-associated protein tau is extensively modified with O-linked N-acetylglucosamine. J Biol Chem 271:28741–28744

Aronov S, Aranda G, Behar L, Ginzburg I (2001) Axonal tau mRNA localization coincides with tau protein in living neuronal cells and depends on axonal targeting signal. J Neurosci 21:6577–6587

Atashi JR, Klinz SG, Ingraham CA, Matten WT, Schachner M, Maness PF (1992) Neural cell adhesion molecules modulate tyrosine phosphorylation of tubulin in nerve growth cone membranes. Neuron 8:831–842

Audesirk G, Cabell L, Kern M (1997) Modulation of neurite branching by phosphorylation in cultured rat hippocampal neurons. Dev Brain Res 102:247–260

Baas PW, Black MM (1990) Individual microtubules in the axon consist of domains that differ in both composition and stability. J Cell Biol 111:495–509

Baas PW, Brown (1997) Slow axonal transport: the polymer transport model. Trends Cell Biol 7:380–384

Bacskai BJ, Hochner B, Mahaut-Smith M, Adams SR, Kaang-B-K, Kandel E, Tsien R (1993) Spatially resolved dynamics of cAMP and protein kinase A subunits in *Aplysia* sensory neurons. Science 260:222–226

Bassell G, Singer RH (1997) mRNA and cytoskeletal filaments. Curr Opin Cell Biol 9:109–115

Binder LI, Frankfurter A, Rebhun LI (1985) The distribution of tau polypeptides in the mammalian central nervous system. J Cell Biol 101:1371–1378

Black MM, Slaughter T, Moshiach S, Obrocka M, Fischer I (1996) Tau is enriched on dynamic microtubules in the distal region of growing axons. J Neurosci 16:3601–3619

Braak H, Braak E, Mandelkow EM (1994) A sequence of cytoskeleton changes related to the formation of neurofibrillary tangles and neuropil threads. Acta Neuropathol (Berl) 87:554–567

Brady RM, Zinkowski RP, Binder LI (1995) Presence of tau in isolated nuclei from human brain. Neurobiol Aging 16:479–486

Bramblett GT, Goedert M, Jakes R, Merrick SE, Trojanowski JQ, Lee VM-Y (1993) Abnormal tau phosphorylation at Ser-396 in Alzheimer's disease recapitulates development and contributes to reduced microtubule binding. Neuron 10:1089–1099

Brandt R, Leger J, Lee G (1995) Interaction of tau with the neural plasma membrane mediated by tau's amino-terminal domain. J Cell Biol 131:1327–1340

Brion JP, Smith C, Couck AM, Gallo JM, Anderton BH (1993) Developmental changes in tau phosphorylation: fetal-type tau is transiently phosphorylated in a manner similar to paired helical filament-tau characteristic of Alzheimer's disease. J Neurochem 61:2071–2080

Brion JP, Octave JN, Couck AM (1994) Distribution of the phosphorylated microtubule-associated protein tau in developing cortical neurons. Neuroscience 63:895–909

Brion JP, Couck AM, Conreur JL, Octave JN (1995) A phosphorylated tau species is transiently present in developing cortical neurons and is not associated with stable microtubules. In: Kosik KS, Christen Y, Selkoe DJ (eds) Alzheimer's disease: lessons from cell biology. Springer, Berlin Heidelberg New York

Butner KA, Kirschner MW (1991) Tau protein binds to microtubules through a flexible array of distributed weak sites. J Cell Biol 115:717–730

Buée L, Delacourte A (2001) Tau phosphorylation. In: Hof PR, Mobbs CV (eds) Functional neurobiology of aging. Academic Press, San Diego, pp 315–332

Buée L, Bussière T, Buée-Scherrer V, Delacourte A, Hof PR (2000) Tau protein isoforms, phosphorylation and role in neurodegenerative disorders. Brain Res Rev 33:95–130

Caceres A, Kosik KS (1990) Inhibition of neurite polarity by tau antisense oligonucleotides in primary cerebellar neurons. Nature 343:461–463

Caceres A, Potrevic S, Kosik KS (1991) The effect of tau antisense oligonucleotides on neurite formation of cultured cerebellar macroneurons. J Neurosci 11:1515–1523

Chen J, Kanai Y, Cowan NJ, Hirokawa N (1992) Projection domains of MAP2 and tau determine spacings between microtubules in dendrites and axons. Nature 360:674–677

Cleveland DW, Hwo SY, Kirschner MW (1977) Purification of tau, a microtubule-associated protein that induces assembly of microtubules from purified tubulin. J Mol Biol 116:207–225

Devoto SH (1990) Neuronal growth cone migration. Experientia 46:916–922

DiTella MC, Feiguin F, Morfini G, Caceres A (1996) MAP1b/tau functional redundancy during laminin-enhanced axonal growth. J Cell Sci 109:467–477

Dotti CG, Banker GA (1987) Experimentally induced alterations in the polarity of developing neurons. Nature 330:254–256

Dotti CG, Sullivan CA, Banker GA (1988) The establishment of polarity by hippocampal neurons in culture. J Neurosci 8:1454–1468

Drubin DG, Feinstein SC, Shooter EM, Kirschner MW (1985) Nerve growth factor-induced neurite outgrowth in PC12 cells involves the coordinate induction of microtubule assembly and assembly-promoting factors. J Cell Biol 101:1799–1807

Dudek SM, Johnson GV (1995) Postnatal changes in serine/threonine protein phosphatases and their association with the microtubules. Dev Brain Res 90:54–61

Ferreira A, Busciglio J, Caceres A (1989) Microtubule formation and neurite growth in cerebellar macroneurons which develop in vitro: evidence for the involvement of the microtubule-associated proteins, MAP1a, HMW-MAP2 and tau. Dev Brain Res 49:215–228

Fritsche J, Reber BF, Schindelholz B, Bandtlow CE (1999) Differential cytoskeletal changes during growth cone collapse in response to hSema III and thrombin. Moll Cell Neurosci 14:398–418

Goedert M, Jakes R (1990) Expression of separate isoforms of human tau protein: correlation with the tau pattern in brain and effects on tubulin polymerization. EMBO J 9:4225–4230

Goedert M, Spillantini MG, Potier MC, Ulrich J, Crowther RA (1989) Cloning and sequencing of the cDNA encoding an isoform of microtubule-associated protein tau containing four tandem repeats: differential expression of tau protein mRNAs in human brain. EMBO J 8:393–399

Goedert M, Jakes R, Crowther RA, Six J, Lübke U, Vandermeerren M, Cras P, Trojanowski JQ, Lee VM-Y (1993) The abnormal phosphorylation of tau protein at Ser-202 in Alzheimer's disease recapitulates phosphorylation during development. Proc Natl Acad Sci USA 90:5066–5070

Goedert M, Jakes R, Vanmechelen E (1995) Monoclonal antibody AT8 recognises tau protein phosphorylated at both serine 202 and threonine 205. Neurosci Lett 189:167–169

Goedert M, Jakes R, Spillantini MG, Hasegawa M, Smith MJ, Crowther RA (1996) Assembly of microtubule-associated protein tau into Alzheimer-like filaments induced by sulphated glycosaminoglycans. Nature 383:550–553

Goode BL, Feinstein SC (1994) Identification of a novel microtubule binding and assembly domain in the developmentally regulated inter-repeat region of tau. J Cell Biol 124:769–782

Gordon-Weeks PR (1987) The cytoskeletons of isolated, neuronal growth cones. Neuroscience 21:977–989

Grundke-Iqbal I, Iqbal K, Quinlan M, Tung Y-C, Zaidi MS, Wisniewski HM (1986) Microtubule-associated protein tau: a component of Alzheimer paired helical filaments. J Biol Chem 261:6084–6089

Haas C, Steiner H (2001) Protofibrils, the unifying toxic molecule of neurodegenerative disorders? Nat Neurosci 4:859–861

Hagestedt T, Lichtenberg B, Wille H, Mandelkow EM, Mandelkow E (1989) Tau proteins become long and stiff upon phosphorylation: correlation between paracrystalline structure and degree of phosphorylation. J Cell Biol 109:1643–1651

Haltiwanger RS, Busby S, Grove K, Li S, Mason D, Medina L, Moloney D, Philipsberg G, Scartozzi R (1997) O-glycosylation of nuclear and cytoplasmic proteins: regulation analogous to phosphorylation? Biochem Biophys Res Commun 231:237–242

Harada A, Oguchi K, Okabe S, Kuno J, Terada S, Ohshima T, Sato-Yoshitake R, Takei Y, Noda T, Hirokawa N (1994) Altered microtubule organization in small-calibre axons of mice lacking tau protein. Nature 369:488–491

Harrison RG (1907) Observations on the living developing nerve fiber. Anat Rec 1:116–118

Heicklen-Klein A, Ginzburg I (2000) Tau promoter confers neuronal specificity and binds Sp1 and AP-2. J Neurochem 75:1408–1418

Heicklen-Klein A, Aronov S, Ginzburg I (2000) Tau promoter activity in neuronally differentiated P19 cells. Brain Res 874:1–9

Hirokawa N, Bloom GS, Vallee RB (1985) Cytoskeletal architecture and immunocytochemical localization of microtubule-associated proteins in regions of axons associated with rapid axonal transport: the IPDN-intoxicated axon as a model system. J Cell Biol 101:1858–1870

Hirokawa N, Funakoshi T, Sato-Harada R, Kanai Y (1996) Selective stabilization of tau in axons and microtubule-associated protein 2C in cell bodies and dendrites contributes to polarized localization of cytoskeletal proteins in mature neurons. J Cell Biol 132:667–679

Ihara Y, Nukina N, Miura R, Ogawara M (1986) Phosphorylated tau protein is integrated into paired helical filaments in Alzheimer's disease. J Biochem 99:1807–1810

Janke C, Beck M, Stahl T, Holzer M, Brauer K, Bigl V, Arendt T (1999) Phylogenetic diversity of the expression of the microtubule-associated protein tau: implications for neurodegenerative disorders. Mol Brain Res 68:119–128

Jenkins SM, Johnson GVW (1997) Phosphorylation of microtubule-associated protein tau on Ser 262 by an embryonic 100-kDa protein kinase. Brain Res 767:305–313

Kamiguchi H, Yoshihara F (2001) The role of endocytic L1 trafficking in polarized adhesion and migration of nerve growth cones. J Neurosci 21:9194–9203

Kanai Y, Takemura R, Ohshima T, Mori H, Ihara Y, Yanagisawa M, Masaki T, Hirokawa N (1989) Expression of multiple tau isoforms and microtubule bundle formation in fibroblasts transfected with a single tau cDNA. J Cell Biol 109:1173–1184

Kanemaru K, Takio K, Miura R, Titani K, Ihara Y (1992) Fetal-type phosphorylation of the tau in paired helical filaments. J Neurochem 58:1667–1675

Kater SB, Mattson MP, Cohan CC, Connor J (1988) Calcium regulation of the neuronal growth cone. Trends Neurosci 11:315–322

Kempf M, Clement A, Faissner A, Lee G, Brandt R (1996) Tau binds to the distal axon early in development of polarity in a microtubule- and microfilament-dependent manner. J Neurosci 16:5583–5592

Kennesey A, Yen S-HC (1993) The extent of phosphorylation of fetal tau is comparable to that of PHF-tau from Alzheimer paired helical filaments. Brain Res 629:40–46

Kosik KS, Joachim CL, Selkoe DJ (1986) Microtubule-associated protein tau is a major antigenic component of paired helical filaments in Alzheimer's disease. Proc Natl Acad Sci USA 83:4044–4048

Kosik KS, Orecchio LD, Bakalis S, Neve RL (1989) Developmentally regulated expression of specific tau sequences. Neuron 2:1389–1397

Lee G, Neve RL, Kosik KS (1989) The microtubule binding domain of tau protein. Neuron 2:1615–1624

Lee G, Newman ST, Gard DL, Band H, Panchamoorthy (1998) Tau interacts with src-family non-receptor tyrosine kinases. J Cell Sci 111:3167–3177

Lin CH, Forscher P (1993) Cytoskeletal remodeling during growth cone-target interactions. J Cell Biol 121:1369–1383

Lindwall G, Cole RD (1984) Phosphorylation affects the ability of tau protein to promote microtubule assembly. J Biol Chem 259:5301–5305

Liu CA, Lee G, Jay DG (1999) Tau is required for neurite outgrowth and growth cone motility of chick sensory neurons. Cell Motil Cytoskel 43:232–242

Lodish H, Baltimore D, Berk A, Zipursky SL, Matsudaira P, Darnell J (1997) Molecular cell biology (eds Freeman WH and Co.). Scientific American Books, New York, pp 1061–1070

Mandell JW, Banker GA (1996) A spatial gradient of tau protein phosphorylation in nascent axons. J Neurosci 16:5727–5740

Matus A, Bernhardt R, Hugh-Jones T (1981) High molecular weight microtubule-associated proteins are preferentially associated with dendritic microtubules in brain. Proc Natl Acad Sci USA 78:3010–3014

Mawal-Dewan M, Henley J, Van de Voorde A, Trojanowski JQ, Lee VMY (1994) The phosphorylation state of tau in the developing rat brain is regulated by phosphoprotein phosphatases. J Biol Chem 269:30981–30987

Mehler MF, Gokhan S (2001) Developmental mechanisms in the pathogenesis of neurodegenerative diseases. Prog Neurobiol 63:337–363

Mesulam M-M (1999) Neuroplasticity failure in Alzheimer's disease: bridging the gap between plaques and tangles. Neuron 24:521–529

Miyata Y, Hoshi M, Nishida E, Minami Y, Sakai H (1986) Binding of MAP2 and tau to the intermediate filament reassembled from neurofilament 70-kDa subunit protein. Its regulation by calmodulin. J Biol Chem 261:13026–13030

Morishima-Kawashima M, Hasegawa M, Takio K, Suzuki M, Titani K, Ihara Y (1995) Proline-directed and nonproline directed phosphorylation of PHF-tau. J Biol Chem 270:823–829

Neve RL, Harris P, Kosik KS, Kurnit DM, Donlon TA (1986) Identification of cDNA clones for the human microtubule-associated protein tau and chromosomal localization of the genes for tau and microtubule associated protein 2. Mol Brain Res 1:271–280

Panda D, Goode BL, Feinstein SC, Wilson L (1995) Kinetic stabilization of microtubule dynamics at steady state by tau and microtubule-binding domains of tau. Biochemistry 34:11117–11127

Ramon y Cajal (1890) A quelle époque apparaissent les extensions des cellules nerveuses de la moëlle épinière du poulet? Anat Anz 5:609–613

Rebhan M, Vacun G, Rösner H (1995) Complementary distribution of tau proteins in different phosphorylation states within growing axons. Neuroreport 6:429–432

Riederer BM, Mourton-Gilles C, Frey P, Delacourte A, Probst A (2001) Differential phosphorylation of tau proteins during kitten brain development and Alzheimer's disease. J Neurocytol 30:145–158

Rösner H, Rebhan M, Vacun G, Vanmechelen E (1995) Developmental expression of tau proteins in the chicken and rat brain: Rapid down-regulation of a paired helical filament epitope in the rat cerebral cortex coincides with the transition from immature to adult tau isoforms. Int J Dev Neurosci 13:607–617

Sato-Yoshitake R, Shiomura Y, Miyasaka H, Hirokawa N (1989) Microtubule-associated protein 1B: Molecular structure, localization, and phosphorylation-dependent expression in developing neurons. Neuron 3:229–238

Schaefer AW, Kamei Y, Kamiguchi H, Wong EV, Rapoport I, Kirchhausen T, Beach CM, Landreth G, Lemmon S, Lemmon V (2002) L1 endocytosis is controlled by a phosphorylation-dephosphorylation cycle stimulated by outside-in signaling by L1. J Cell Biol 157:1223–1232

Selden SC, Pollard TD (1983) Phosphorylation of microtubule-associated proteins regulates their interaction with actin filaments. J Biol Chem 258:7064–7071

Šimić G, Gnjidić M, Kostović I (1998) Cytoskeletal changes as an alternative view on pathogenesis of Alzheimer's disease. Period Biol 100:165-173

Slaughter T, Wang J, Black MM (1997) Microtubule transport from the cell body into the axons of growing neurons. J Neurosci 17:5807–5819

Smith CJ, Anderton BH, Davis DR, Gallo JM (1995) Tau isoform expression and phosphorylation state during differentiation of cultured neuronal cells. FEBS Lett 375:243–248

Snow DM, Mullins N, Hynds DL (2001) Nervous system-derived chondroitin-sulfate proteoglycans regulate growth cone morphology and inhibit neurite outgrowth: a light, epifluorescence, and electron microscopic study. Microsc Res Tech 54:273–286

Szaro B, Lee V, Gainer H (1989) Spatial and temporal expression of phosphorylated and non-phosphorylated forms of the neurofilament proteins in the developing nervous system of *Xenopus laevis*. Dev Brain Res 48:87–103

Szendrei GI, Lee VMY, Otvos L (1993) Recognition of the minimal epitope of monoclonal antibody Tau-1 depends upon the presence of a phosphate group but not its location. J Neurosci Res 34:243–249

Takahashi M, Tsujioka Y, Yamada T, Tsuboi Y, Okada H, Yamamoto T, Liposits Z (1999) Glycosylation of microtubule-associated protein tau in Alzheimer's disease brain. Acta Neuropathol (Berl) 97:635–641

Takei Y, Teng J, Harada A, Hirokawa N (2000) Defects in axonal elongation and neuronal migration in mice with disrupted tau and map1b genes. J Cell Biol 150:989–1000

Tanaka E, Sabry J (1995) Making connection: cytoskeletal rearrangements during growth cone guidance. Cell 83:171–176

Tint I, Slaughter T, Fischer I, Black MM (1998) Acute inactivation of tau has no effect on dynamic of microtubules in growing axons of cultured sympathetic neurons. J Neurosci 18:8660–8673

Uberti D, Rizzini C, Spano P-F, Memo M (1997) Characterization of tau proteins in human neuroblastoma SH-SY5Y cell line. Neurosci Lett 235:149–153

Vickers JC, Dickson TC, Adlard PA, Saunders HL, King CE, McCormack G (2000) The cause of neuronal degeneration in Alzheimer's disease. Prog Neurobiol 60:139–165

Wang JZ, Grundke-Iqbal I, Iqbal K (1996) Glycosylation of microtubule-associated protein tau: an abnormal posttranslational modification in Alzheimer's disease. Nat Med 2:871–875

Watanabe A, Hasegawa M, Suzuki M, Takio K, Morishima-Kawashima M, Titani K, Arai T, Kosik KS, Ihara Y (1993) In vivo phosphorylation sites in fetal and adult rat tau. J Biol Chem 268:25712–25717

Williamson T, Gordon-Weeks PR, Schachner M, Taylor J (1996) Microtubule reorganization is obligatory for growth cone turning. Proc Natl Acad Sci USA 93:15221–15226

Wood JG, Mirra SS, Pollock NJ, Binder LI (1986) Neurofibrillary tangles of Alzheimer disease share antigenic determinants with the axonal microtubule-associated protein tau. Proc Natl Acad Sci USA 83:1–4

Yamada KM, Spooner BS, Wessells NK (1971) Ultrastructure and function of growth cones and axons in cultured nerve cells. J Cell Biol 49:614–635

Zmuda JF, Rivas RJ (2000) Actin disruption alters the localization of tau in the growth cones of cerebellar granule neurons. J Cell Sci 113:2797–2809

Developmental Expression and Possible Roles of Gangliosides in Brain Development

H. Rösner

1 Introduction

Gangliosides are sialic acid containing glycosphingolipids structurally defined as hematosides, lacto-, neolacto-, globo-, and ganglio-series gangliosides (Ledeen 1978; Ledeen and Yu 1982; Wiegandt 1982, 1995; Yu 1994). Up to date, about 100 different ganglioside structures have been identified in neuronal and extraneural tissues (Yu 1994). Gangliosides are principle membrane constituents of vertebrate cells, synthesized by step-wise addition of carbohydrate moieties to ceramide during its transfer from the ER through the *cis*- and *trans*-Golgi compartments (for review see Tettamanti et al. 1987) and degraded via lysosomal pathways (Sandhoff et al. 1987). Their functional significance is still obscure. However, numerous experimental data suggest that apart from their basic function as structural membrane components per se, these glycolipids play an important role in signal transduction (Hakomori 1981) including cell/cell (Schnaar et al. 1998) and cell/extracellular matrix interactions (Cheresh et al. 1986). Three modes of action whereby gangliosides and their degradation products affect signal transduction have been proposed: (1) modulation of growth factor/hormone and integrin receptors, (2) modulation of cell cycle and proliferation by ganglioside derivatives, and (3) induction of signalling, together with transducer molecules by gangliosides organized in membrane microdomains (Kojima and Hakomori 1991; Hakomori et al. 1998; Probstmeier and Pesheva 1999; Prinetti et al. 2000). The mature brain contains about 20-fold higher amounts of gangliosides than external tissues (Ledeen and Yu 1982). Neuronal membranes are especially enriched in ganglio-series gangliosides synthesized along different pathways (Fig. 1). While a- and b-gangliosides predominate in the adult avian and mammalian brain, in fish nervous tissue b- and mainly c-gangliosides are abundant

Institute of Zoology, University of Hohenheim-Stuttgart, Garbenstr. 30, 70593 Stuttgart, Germany

Progress in Molecular and Subcellular Biology, Vol. 32
Ivica Kostović (Ed.)

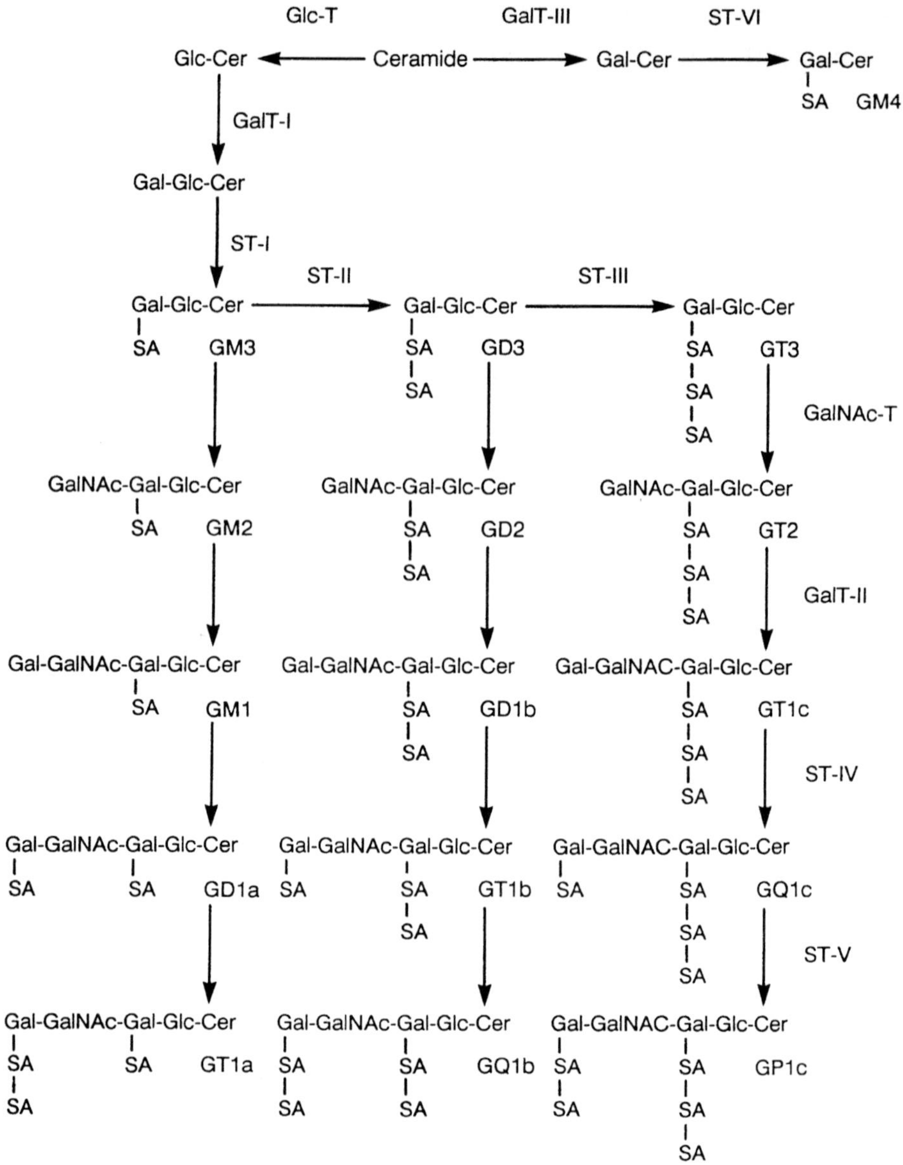

Fig. 1. Biosynthesis pathways of gangliosides; nomenclature of gangliosides; *ST* Sialyltransferase; *GalT* galactosyltransferase; *Glc-T* glycosyltransferase; *GalNAc-T* N-acetylgalactosaminyltransferase

showing a high degree of diversity (Hilbig and Rahmann 1987). In addition, striking and neural cell-type-specific regional differences of the ganglioside composition have been established in the mammalian (including human) brain (Suzuki 1965; Rösner 1977; Kracun et al. 1984; Kotani et al. 1993, 1994, 1995). Unlike neurons, the ganglioside pattern of astrocytes resembles that of most nonneural cells containing much less gangliosides and a prevalence of GM3 and GD3 (Robert et al. 1977; Sbashing-Agler et al. 1988; Asou et al. 1989). Likewise, oligodendroglia and central myelin seem to express less sialylated compounds, mainly GM1 and GM4 in addition to GM3 and GD3 (Cochran et al. 1982; Kim et al. 1986). Schwann cells and peripheral myelin were shown to be enriched in ganglioside LM1 (Chou et al. 1982). Taking into consideration different sialic acid structures (n-/o-acetyl, n-glycolyl), as well as lactonization, fucosylation and different ceramide moieties (C18-C24) one can expect a high degree of cell- and species-specific variability of the ganglioside expression.

Beginning with the report of Suzuki (1965), numerous studies demonstrated that the expression of nervous system gangliosides is a spatially and temporally regulated process with some variation between species.

The aim of this presentation is to confirm that, in spite of striking cell- and species-specific differences, there are some general principles of developmental changes of brain ganglioside expression common to all higher vertebrates, including humans.

2
Gangliosides and Brain Development

Indication for the assumption that gangliosides may play a crucial role in nervous tissue development goes back to the discovery of these substances in the grey matter of normal and pathological human brains by Klenk (1942) and the later detection of inherited disorders of ganglioside metabolism leading to abnormal brain development and neuronal dysfunction (Fishman and Brady 1976; Sandhoff and Christomanou 1979). Since then, numerous in vivo and in vitro studies using biochemical, enzymatic, immunological and genetic approaches confirmed that neuronal (and glial) ganglioside expression is strictly developmentally regulated in the sense that changes in ganglioside concentration and composition coincide with the progress of neuronal (and glial) differentiation and maturation.

Brain development is one of the most complex examples of morphogenesis. It can be considered as divided into epochs defined by more or less irreversible processes (Fig. 2): (1) neural induction, which leads to the formation of the neural tube and its anterior-posterior pattering; (2) proliferation of neuronal and glial progenitor cells, some of the latter differentiating to ependymal and radial glial cells; (3) withdrawal of neuroblasts from their last mitotic cycle and migration of new born neurons involving neuron/radial glial and neuronal interactions; (4) outgrowth of axons, fibre tract mapping, arborization of dendrites, synaptogenesis involving neuronal apoptosis; (5) enhanced proliferation of glial progenitor cells, differentiation to astroglial and oligodendroglial cells and myelination; (6) continuing histogenetic rearrangement involving later born microneurons, in which increasing function determines final connections; period of the mature brain characterized by long-lasting synaptic plasticity; (7) aging and slowly increasing neuronal degeneration. The extent and duration of these epochs differ among species and more or less overlap. Nevertheless, clear-cut and sometimes irreversible events characteristic of different epochs can be seen to occur at the cellular and molecular level.

2.1 Neural Induction and Tube Formation

Only a few studies provide data concerning ganglioside expression during this very early stage of brain development. Felding-Habermann et al. (1986) found by radiolabeling that the "prebrain" – chicken embryo synthesizes small amounts of GD3 and several other glycosphingolipids different from ganglioseries-gangliosides. Thierfelder et al. (1992) reported an immunostaining of GD3 by mAb R24 of the whole chicken blastoderm of stages 4–5 according to Hamburger and Hamilton (1951). At stages 12–13, the staining was restricted to the endophyllic crescent and cranial part of the notochord. This observation corresponds to results by Rösner et al. (1985a, 1992) who found up to around stage 8 of chicken development, an immuno-expression of GD3 which was restricted to heavily proliferating cells of mainly mesodermal origin (Fig. 3). At this early developmental stage neural and glial progenitor cells of the forming neural groove and tube, however, did not react with mAb R24 or other antibodies directed at GD2, GM1, GD1b or c-gangliosides.

These data suggest that during neural tube formation and anterior-posterior pattering, lasting in the chicken embryo up to around stage 8 (26–29 h), the progenitor cells of glia and neurons neither express GD3 nor higher ganglio-series gangliosides in amounts detectable by specific antibodies.

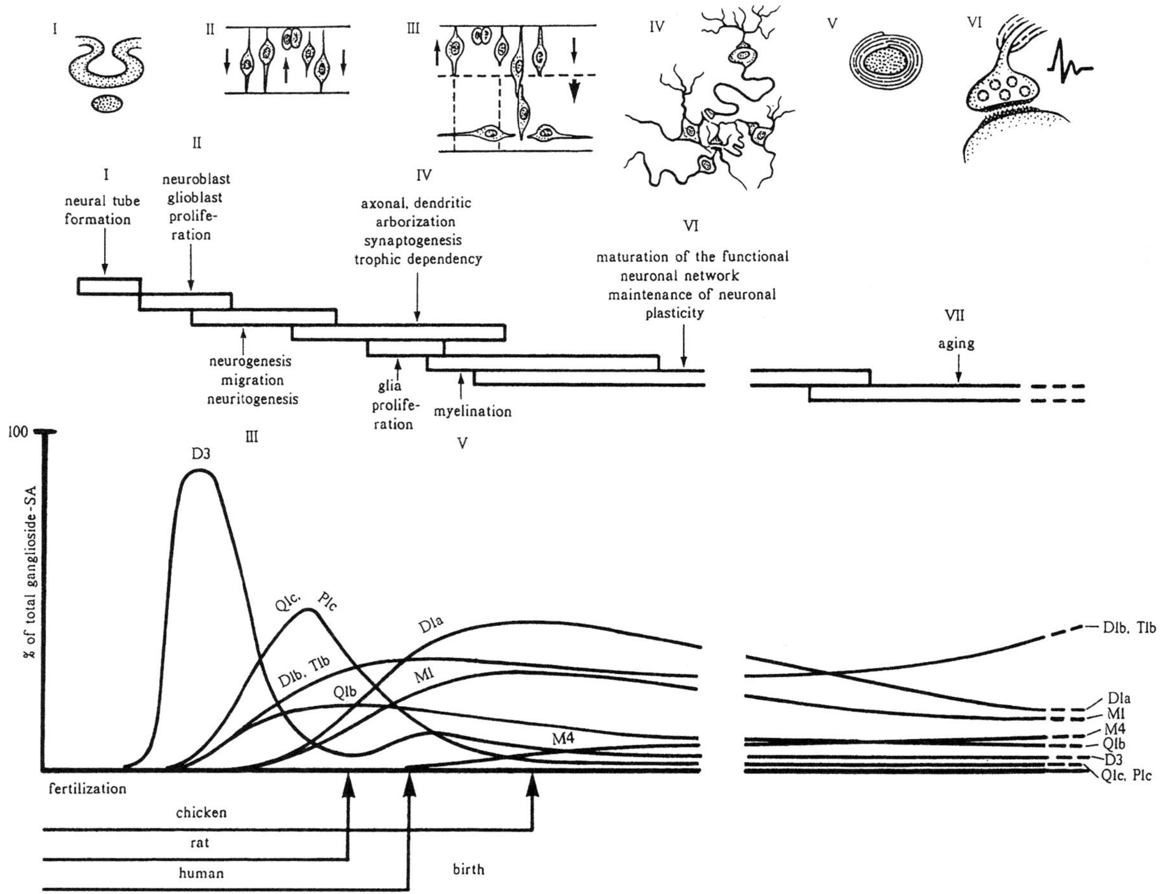

Fig. 2. Generalized presentation of the relative developmental and aging profiles of brain gangliosides of higher vertebrates in correlation with the progress of morphological and functional maturation (*I–VII*); birth is indicated by *arrows* for chicken, rat and human

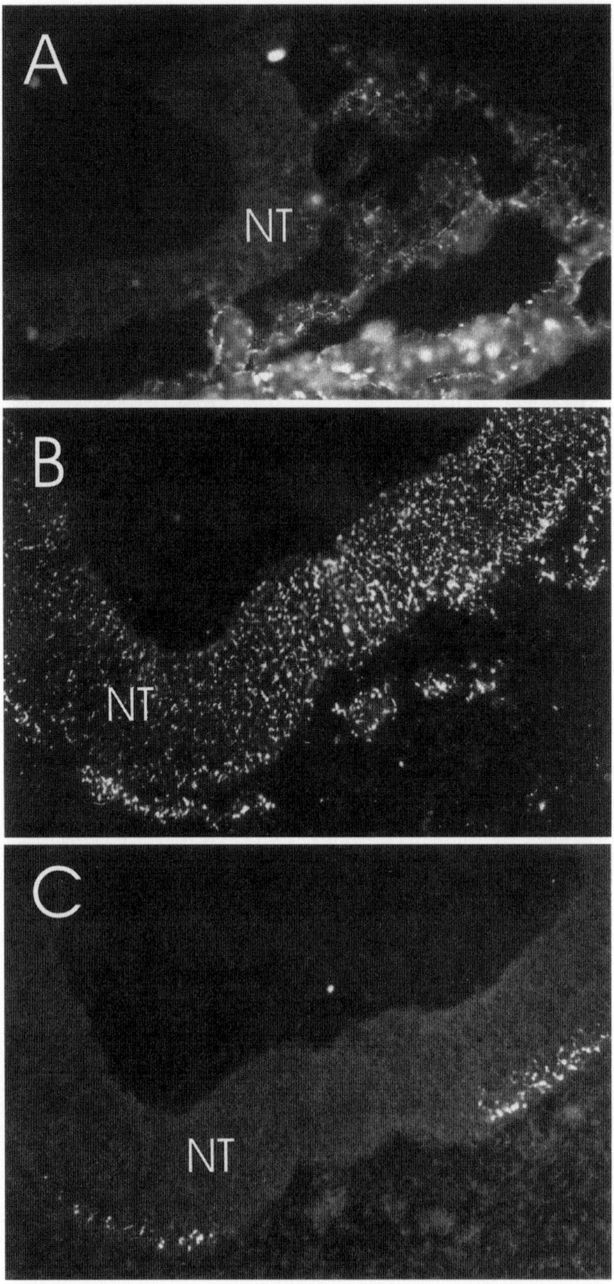

Fig. 3A–C. Immuno-expression of GD3 and c-gangliosides in chicken neuroepithelium detected by the mAbs R24 and Q211, respectively. **A** In contrast to proliferating mesodermal cells, the forming neural tube (*NT*; 26–29 h, stage 8) does not express GD3 (or any other gangliosides). **B** High expression of GD3 in the midbrain neural tube (*NT*) of 2.5-day-old embryos. **C** First expression of c-gangliosides by "new born" neurons located at the floor of the midbrain neural tube (*NT*)

2.2
Proliferation of Neural and Glial Progenitor Cells

Numerous biochemical studies indicated that GD3 is the predominant ganglioside of the early, immature nervous system of birds and mammals (including humans), but a minor glycolipid of the mature brain (Dreyfus et al. 1975; Rösner 1977, 1980, 1982; Yavin and Yavin 1979; Hilbig et al. 1982; Goldmann et al. 1984; Rösner et al. 1985a; Schaal et al. 1985; Seyfried and Yu 1985; Constantine-Paton et al. 1986; Felding-Habermann et al. 1986; Maccioni et al. 1989). More detailed information concerning the spatiotemporal cellular expression of GD3 was obtained by means of immunohistochemical staining using specific mAbs (Rösner et al. 1985a, 1992).

Thus, we observed that the first expression of GD3 in chicken neuroepithelium occurs in the anterior prosencephalic vesicle of stages 9–10 (1.5 days). During the next 3 days (up to stages 25–27), immuno-expression of GD3 increases from rostral to caudal in all regions of the forming brain vesicles and spinal cord in parallel with the spatio-temporal increase in proliferative activity (Figs. 3, 4). Likewise, in the retina (Fig. 4) and otic anlage, GD3 immunolabeling follows the spatio-temporal gradient of proliferation from the center to the periphery. At the same time, migrating neuronal crest and mitotic nonneural cells (e.g. lens, endothelia of forming blood vessels) express GD3 profusely. Similar results have been revealed for the embryonic rat brain of comparable stages demonstrating an intense immuno-expression of GD3 by growing endothelia, by immature neuro- and glioblasts of the ventricular and subventricular layers of the prenatal forebrain (Fig. 4; Rösner et al. 1986, 1988a, b), by germinal layers of the rat cerebellum (Schaal et al. 1985; Goldmann et al. 1984), dentate gyrus, and hippocampus (Goldmann et al. 1984), as well as by the primordium of the 5-week-old human brain and the human hippocampus and pallium at 17 weeks of gestation (Kracun et al. 1992). It is important to note that at this early stage of development mAbs directed at tetraosyl gangliosides as GD1b, or c-polysialogangliosides never stained the ventricular and subventricular germinal zones, neither in embryonic chicken nor in rat brain (Rösner et al. 1985a, 1986, 1988a, b; Figs. 3, 4). These "higher" gangliosides appear for the first time on the cell surface of "new born" neurons (see below). Biochemical data (Rösner 1980, 1982) showed that in the chicken neuroepithelium of E5 consisting of more than 80% of immature cells, GT3 is also abundant. Although enzymatic data of these early stages are lacking, the expression of GD3 and GT3 suggest a high activity of ST-II and ST-III (Fig. 1) of neuronal and glial precursors cells.

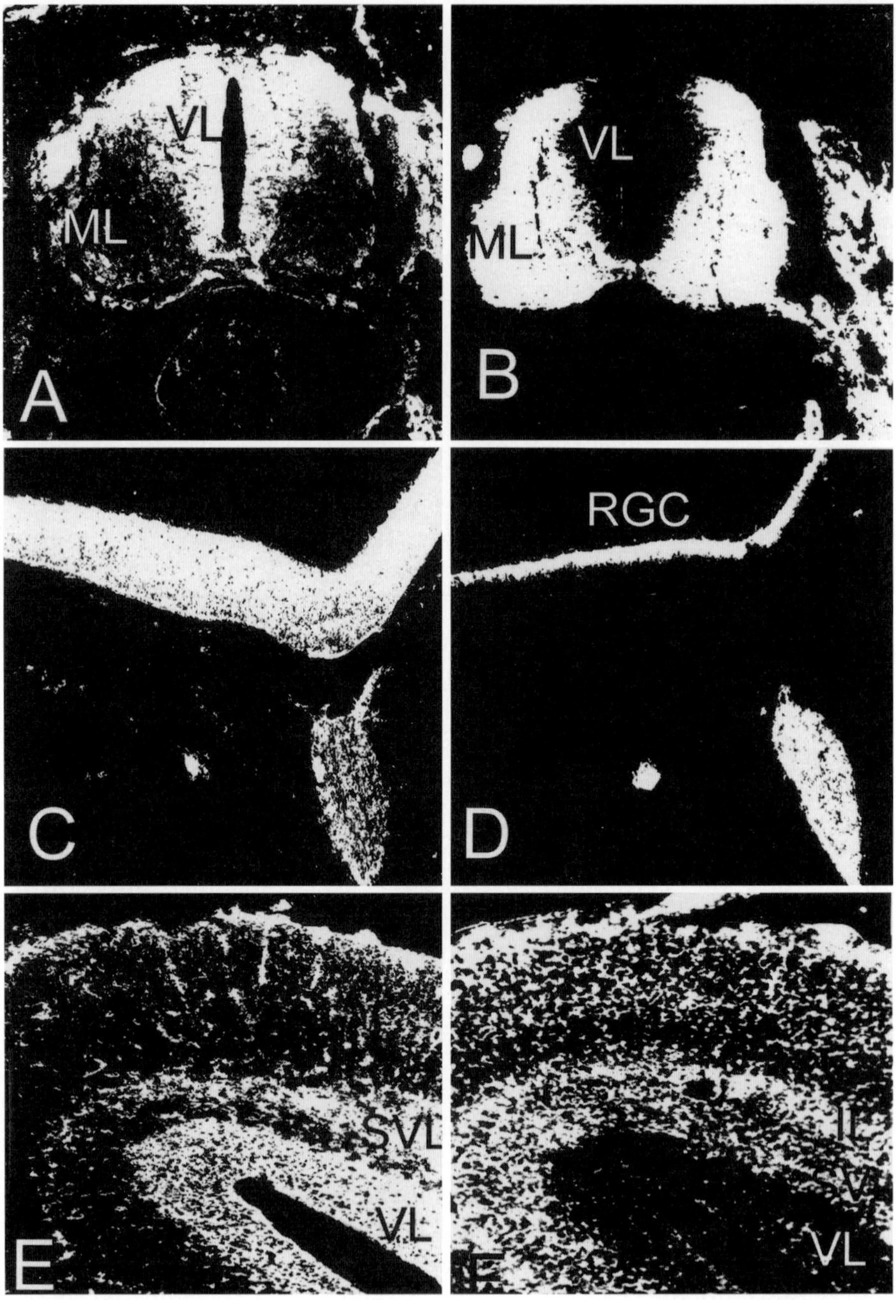
VL
ML
A
VL
ML
B
C
RGC
D
SVL
VL
E
II
V
VL
F

From the above data the general conclusion seems to be justified that heavily proliferating embryonic neuroectodermal (neuronal and glial) progenitor cells, neural crest as well as nonneural stem and progenitor cells are characterized by an intense expression of ganglioside GD3.

In this respect, these mitotically active and undifferentiated embryonic cells behave in a similar way to reactive glia (Levine et al. 1986; Yu et al. 1988) and many transformed neuroectodermal cells as melanomas (Pukel et al. 1982), gliomas and astrocytomas (Eto and Shinoda 1982), meningiomas (Fredmann et al. 1990) and neuroblastomas (Rebhan et al. 1994). Interestingly, induction of ST-II activity and accretion of GD3 in rat fibroblasts by transfection with a corresponding c-myc-tagged DNA, correlated with the ability of anchorage-independent growth known to be a reliable feature of tumorigenicity in vitro (Nakaishi et al. 1988). At this point, it should be noted that GD3 is, besides GM3, the main ganglioside of many extraneural cells and glial cells. Its concentration, however, is much lower in these cells than in embryonic, reactive or transformed cells.

2.3 Neurogenesis and Neuron Migration

Proceeding in brain development, neuroblasts withdraw from their last mitotic cycle to become postmitotic and migrate from the ventricular into the forming mantle layers. Numerous biochemical studies of different areas of the developing fish, avian and mammalian (including human) brain revealed that during this developmental period there is a relative decrease of GD3 in favor of an accretion of higher ganglioseries – gangliosides with a complete tetraose-moiety and up to five sialic acid residues (Suzuki 1965; Merat and Dickerson 1973; Vanier et al. 1973; Dreyfus et al. 1975; Rösner 1975, 1980, 1982; Yavin

Fig. 4A–F. Immuno-expression of GD3 and c-gangliosides in the embryonic chicken and rat brain detected by the mAbs R24 and Q211, respectively. **A** E5 chicken spinal cord showing intense expression of GD3 in the ventricular layer (*VL*) and dorsal root. **B** Expression of c-gangliosides by postmitotic neurons of the E5 chicken spinal cord, spinal ganglia and dorsal root ganglia. **C** GD3 expression in the whole retina and optic stalk of E5 chicken. **D** Expression of c-gangliosides restricted to retinal ganglion cells (*RGC*) in the E5 chicken retina. **E** Intense expression of GD3 in the ventricular layer (*VL*), subventricular layer (*SVL*) and endothelia of forming blood vessel in the E17 rat cerebrum. **F** Expression of c-gangliosides by postmitotic neurons of the subventricular layer (*SVL*), intermediate layer (*IL*) and cortical plate in the E17 rat cerebrum

and Yavin 1979; Hilbig et al. 1982; Goldmann et al. 1984; Maccioni et al. 1984, 1989; Rösner et al. 1985a, 1988a, b; Schaal et al. 1985; Seybold and Rahmann 1985; Greis and Rösner 1990a, b). Similar changes have been found in cell and tissue cultures from embryonic chick retina (Landa and Moscona 1985; Panzetta et al. 1987) and optic tectum (Engel et al. 1979) and in rat primary cultures (Thangnipon and Balazs 1992). As a further general phenomenon, it was observed that developmental expression of b-pathway gangliosides precedes the accretion of a-pathway gangliosides. In addition, we found in embryonic chicken brain several "novel" highly sialylated gangliosides to be expressed in parallel with b-gangliosides (Fig. 5; Rösner 1980, 1982). Sialic acid to sphingosine ratios, tlc identification of enzyme-digested derivates and FAB-spectrometric identification of the main component as GP1c (Rösner 1981; Rösner et al. 1985b) established that these gangliosides represent glycosphingolipids of the c-pathway characterized by three sialic acid residues linked to the inner galactose (Figs. 1, 5). C-gangliosides were first identified in goldfish brain (Ishizuka and Wiegandt 1972) and are major gangliosides of the brains of bony fishes and rajiform elasmobranchs. They are synthesized via a separate pathway (Fig. 1; Yu and Ando 1980). The presence of a whole series of c-gangliosides (GT3, GT2, GT1c, GQ1c, GP1c, and GH) in embryonic chicken brain has been confirmed by immuno-tlc by means of the mAbs Q211 (Fig. 5; Rösner et al. 1985a) and M6704 and M7103 (Hirabayashi et al. 1988a). Subsequently, some of these gangliosides were detected in the perinatal rat cerebrum (Rösner et al. 1988b) and retina (Daniotti et al. 1992), the fetal human brain (Greis and Rösner 1990a, b; Kracun et al. 1992) and in the adult brains of turtles (Greis and Rösner 1990a; Kracun et al. 1992; Letinic et al. 1998) and bovine (Hirabayashi et al. 1988a, b, 1990).

Immunohistochemical staining by means of mAb Q211 of cryosections from different areas of the embryonic chicken revealed c-polysialogangliosides to be neuron-specific. Their first expression was found to coincide with differentiation of mitotic neuroblasts to post-mitotic young neurons in both the central and peripheral nervous system in vivo (Figs. 3, 4; Rösner et al. 1985a) and in vitro (Rohrer et al. 1985; Rösner et al. 1992). Immunostaining of comparable cryosections with mAbs to GD1b showed that at the same time new-born neurons express gangliosides of the b-pathway (Fig. 5B). These b- and c-ganglioside-positive neurons leave the germinal layers and migrate into the mantle layers (Rösner et al. 1992). At the same time, migrating young neurons show a decreased expression of GD3 (Figs. 3, 4 and 5; Rösner et al. 1985a). Comparable immunohistochemical data from the embryonic

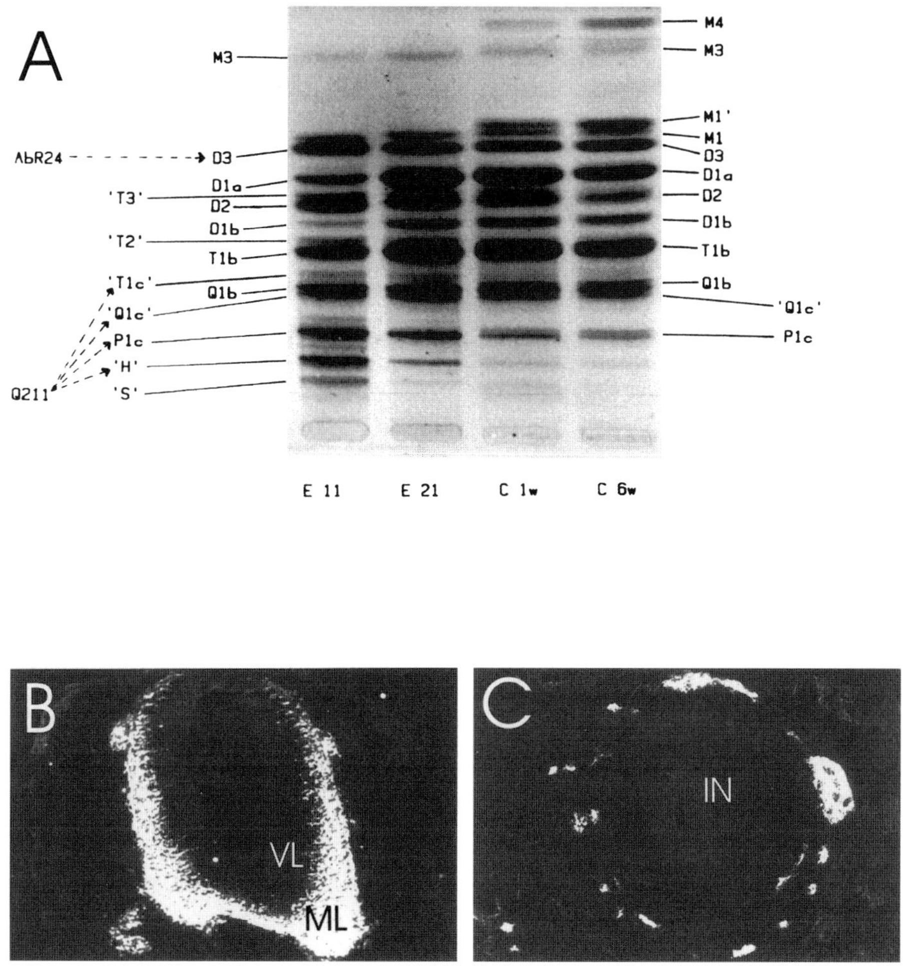

Fig. 5. **A** Ganglioside pattern of the chicken optic lobes of embryonic day 11 (*E11*) to 6 weeks (*6 W*) after hatching; decrease in the c-pathway gangliosides *T3*, *T2*, *T1c*, *Q1c*, *P1c*, *H*, and *S*; increase in D1b, T1b, D1a (neurons) and M1′, M4 (myelin). **B** First expression of GD1b (detected by mAb GGR12) in postmitotic migrating neurons of the E6 chicken spinal cord. **C** Expression of c-gangliosides (detected by mAb Q211) in postmitotic neural crest derivatives of embryonic chicken forming the intramural nervous system; *IN* intestine

mammalian nervous system are still scattered. However, results from primary cultures confirmed that the expression tetraosyl gangliosides of the b-pathway is restricted to neurons and does not occur in mitotic neuroblasts (Thangnipon and Balazs 1992). Furthermore, in the prenatal rat embryonic brain, we found increasing immuno-expression of tetraosyl-c gangliosides by new born cortical neurons and by thalamic neurons forming the thalamo-cortical/hippocampal tract (Rösner et al. 1988b; Greis and Rösner 1990a, b). In contrast to the mitotic neuronal (and glial) progenitor cells of the ventricular and subventricular layer, these neurons no longer expressed GD3 in immunodetectable amounts.

In summary, the biochemical and immunohistological data obtained from embryonic chicken and mammalian brain showed that the transition of mitotic neuroblasts to postmitotic neurons is characterized by the first expression of tetraosyl gangliosides of the c- and b-pathways followed by the synthesis a-pathway gangliosides. At the same time, in most neuron types if not all (see below), the cell-surface expression of GD3 (as well as of GD2, GT3 and GT2) decreases to a level that is no longer detectable by antiganglioside antibodies.

From the above data, it must be concluded that neurogenesis in the vertebrate brain is associated with a rapid maturation of the entire Golgi complex including mechanisms transferring GD3 and GT3 from the *cis*- through the *trans*-Golgi compartments where glycosylation is completed. According to the accepted concept of ganglioside biosynthesis (Pohlentz et al. 1988; van Echten et al. 1990; Iber et al. 1990), the maintenance of high activities of sialyltransferases II and III (ST-II, ST-III; Fig. 1) compared to GalNAc-T, and a rise of Gal T-II as well as ST-IV and ST-V could explain a drop in GD3 and GT3 due to an increased synthesis of tetraosyl b- and c-gangliosides, while formation of gangliosides (via GM2) would remain low. In this model, in addition to a Golgi transport-mediated regulation, Gal T-II would be a key enzyme in "opening" the pathways to tetraosyl b- and c-gangliosides. Enzymatic studies have been performed with chicken retina of E7 in vivo (Maccioni et al. 1989) and in vitro (Landa et al. 1984), the youngest stage investigated so far. These data demonstrate a higher activity of ST-II compared to Gal T-II and GalNAc-T, which fits the above model well. However, as the E7-retina contains only about 30% neurons, a discrimination between immature cells and neurons with respect to enzyme activities was not possible. At later developmental stages, ST-II activity of chicken retina was found to decrease and that of GalNAc-T and GAL T-II to increase (Panzetta et al. 1980; Landa et al. 1984; Maccioni et al. 1989). In embryonic rat brain, ST-II has been reported to remain on a high level up to E14 (Maccioni et al. 1984), respectively up to E18 (Yu et al. 1988).

The latter data suggested that a neurogenesis-associated decrease in the expression of GD3 may not be shared by all neuron types. In fact, cerebellar Purkinje cells are reported to maintain an intense expression of GD3 on their dendritic trees up to the adult stage (Reynolds and Wilkin 1988). The same was shown for differentiated rat retinal neurons

(Daniotti et al. 1990, 1991, 1992). Furthermore, alkali-labile (9-O-acetyl) forms of GD3 (and GD2), which are immuno-recognized by the Jones antibody, have been found to be abundant in the perinatal rat retina (Constantine-Paton et al. 1986). The Jones antigen(s) seem to be expressed by glial cells and neurons (Mendez-Otero and Constantine-Paton 1990) and are found to be developmentally regulated elsewhere in the rat and mouse brain in correlation with the periods of cell migration (Mendez-Otero et al. 1988; Schlosshauer et al. 1988). Furthermore, a developmental regulation in the central nervous system of O-acetylated minor gangliosides detected by the mAb 493D4 has been reported (Zhang et al. 1997).

2.4
Outgrowth of Axons and Fibre Tract Mapping, Dentritogenesis, Synaptogenesis and Apoptosis

After reaching their final destinations by migration, young neurons form (often long) axonal processes which may arborize in appropriate target areas, followed by outgrowth and arborization of dendrites, synaptogenesis and selective neuronal apoptosis. During this period of the so-called growth spurt, different neuron types strikingly increase in size (increase in membrane surface-to-volume ratio) and differentiate with respect to morphology, growth factor-dependency and transmitter specificity. During the period of growth spurt, the duration of which differs remarkably among avian and mammalian species, again striking changes of ganglioside expression are established. Gross biochemical analysis of different brain areas revealed a decrease in c-gangliosides in the pre-hatched chicken and early postnatal rat brain (Fig. 2; Rösner 1982; Rösner et al. 1988a, b; Greis and Rösner 1990a, b). In contrast, during this period, in both the avian and mammalian brain, there is an area-specific accretion of gangliosides of the b-pathway (GD1b, GT1b, GQ1b) followed by an even more striking area-specific accretion of a-pathway gangliosides (GM1, GD1a, GT1a; Fig. 2), leading to a several-fold increase in total ganglioside content based on brain fresh weight or total protein. This increase in tetraosyl gangliosides of the a- and b-pathway was shown to be a general feature of the developing avian and mammalian brain during "growth spurt" and synaptogenesis and is probably due to an increased proportion of neuronal membranes (Suzuki 1965; Merat and Dickerson 1973; Vanier et al. 1973; Rösner 1975, 1977, 1980, 1982; Mannson et al. 1978; Dreyfus et al. 1980; Irwin et al. 1980; Panzetta et al. 1980, 1987; Hilbig et al. 1982, 1984; Bouvier and Seyfried 1989; Landa et al. 1984; Maccioni et al. 1984, 1989; Kracun et al.

1986; Seyfried 1987; Yu et al. 1988; Svennerholm et al. 1989; Sonnino et al. 1990). Among the minor gangliosides, GM2 was also found to increase several-fold (Rösner 1982), probably specifically due to dendritic arborization and growth (Zervas and Walkley 1999; Walkley et al. 1995, 1998).

On the enzymatic level (Fig. 1), these changes could be caused by a decrease in the activities of ST-II and ST-III and an increase in GalNAc-T activity, leading to a shift in favor of GM2 and subsequently of GM1 and GD1a. In fact, a decrease in ST-II activity concomitantly with a rise in GalNAc-T activity has been shown to occur in chicken retina (Panzetta et al. 1980; Landa et al. 1984; Maccioni et al. 1989) and embryonic rat brain (Maccioni et al. 1984; Yu et al. 1988). The significance of the relative activities of ST-II vs. GalNAc-T in regulation of the ratios of both lactosyl to gangliotetraosyl gangliosides and a- to b-gangliosides was confirmed by Daniotti et al. (1991). These authors demonstrated that, unlike chicken, the rat retina maintains a high ST-II/GalNAc-T activity ratio up to adulthood, resulting in a much lower ratio of a- to b-gangliosides and a much higher content of GD3. Moreover, a shift in synthesis from b- to a-gangliosides and vice versa was shown in primary cultures of murine cerebellar cells to depend directly on the activities of ST-II and GalNAc-T, which could be experimentally modulated by a change of pH (Iber et al. 1990). Modulation of membrane flow by drugs, a lowering of ambient temperature or application of metabolic inhibitors (van Echten and Sandhoff 1989; van Echten et al. 1990) further supports the concept that sialylation of GM3 to GD3 (ST-II), of GD3 to GT3 (ST-III), and glycosylation of GM3 to GM2 (GalNAc-T) are the most important regulatory steps in promoting ganglioside synthesis through either the b- (and c-) or the a-pathway. A transport-mediated regulation of the later steps of glycosylation within the *trans*-Golgi complex seems to be unlikely, but cannot be excluded. Another open question is whether the enzyme activities are regulated on the transcriptional or post-transcriptional level or on both. Data obtained from oncogene-transfected cell lines suggest that both regulatory mechanisms are effective (Nagai et al. 1987).

2.5 Myelination

Turning back to the development of the avian and mammalian brain, the period of growth spurt and synaptogenesis is immediately followed by myelination (Fig. 2). Concerning gangliosides, this period is characterized by a second rise of GD3, probably due to an increased proliferation of oligodendroglia cells (Rösner 1982) and an accretion of GM4 (Fig. 2; Rösner 1982), a minor ganglioside shown to be enriched in oligodendroglia and central myelin (Cochran et al. 1982). In the chicken brain, the increase in GM4 is paralleled by a rise of a neuraminidase-resistant isotype of GM1 (Rösner 1982), which is probably specific for avian central myelin. During peripheral myelination, there is an increasing accretion of mainly lacto- and neolacto-series gangliosides including LM1 (Chou et al. 1982; Ogawa-Goto and Abe 1998).

2.6 Structural and Functional Maturation

Proceeding in development, in addition to species-specific, area-specific changes of the brain ganglioside composition also occur depending on the time course and degree of morphological and functional maturation. Thus, cerebellar gangliosides change later in development as compared to the cerebrum due to a later morphological and functional differentiation. In addition, the high degree of area-specific differences in the ganglioside composition of the adult human brain (Fig. 6; Suzuki 1965; Kracun et al. 1983, 1984) suggests local changes during final morphological and functional maturation leading to a cell type-specific final ganglioside expression. In fact, up to 4 months after birth, the human frontal and occipital cortex and hippocampus have similar development profiles of individual gangliosides (Kracun et al. 1986, 1991, 1992). Thereafter, however, there is a drop in GD1a in the occipital cortex, leading to a predominance of b-gangliosides, especially in the visual cortex. In contrast, in the frontal cortex, there is only a small decrease and in the hippocampus no decrease in a-gangliosides, which resulted in a final prevalence of GD1a and in GM1 in these brain areas (Fig. 6). Using a set of specific antiganglioside antibodies, Kotani et al. (1993, 1994) clearly confirmed by immunohistochemical staining a high degree of area- and cell-specific expression of gangliosides in the adult rat central nervous system. At this point, it is important to note that, in addition to the main ganglio-series gangliosides, many other glycosphingolipids have been described to be expressed as minor fractions in the mature brain more or less transiently throughout development. Chol-1-family gangliosides have been shown to be peculiar to certain cholinergic neurons (Richardson et al. 1982; Derrington and Borroni 1990). The lacto-series gangliosides 3′-LM1, 3′-iso LM1 and LD1 were demonstrated to constitute a significant portion with different profiles in fetal human brain. Fucosyl-GM1 was shown to increase in postnatal human brain (Svennerholm et al. 1989). In the chicken brain, the gangliosides

Neocortex
- Prefrontal (Pf)
- Frontal (F)
- Orbifrontal (Of)
- Area subcallosa (As)
- Gyrus precentralis (gP)
- Gyrus cinguli (gC)
- Temporal (T)
- Insular (I)
- Parietal (Pr)
- Occipital (Oc)
- Striatal (Sr)

Cerebellum
- Cortex (Cc)
- Vermis

Paleocortex
- S. perforata ant. (spa)
- Cortex periamygdalaris (Pc)
- N. caudatus caput (NCc)
- N. caudatus corpus (NCcr)
- N. accumbens (Nac)
- Putamen (P)
- Globus pallidus (GP)
- Amygdala (A)
- N. basalis Maynert (NBM)

Archicortex
- Hippocampus CA2 (h)
- Hippocampus CA3 (h)
- Hippocampus CA4 (h)
- Hippocampus proper (h)
- Periarchicortex (Pea)

Diencephalon
- Thalamus (Th)
- Thalamus pulvinar (Thp)
- Hypothalamus (Hy)
- Corpora mamillaria (CM)
- Epiphysis (E)

Mesencephalon
- Colliculi superiores (CS)
- Colliculi inferiores (CIn)
- Substantia nigra (Sn)
- Nucleus ruber (Nr)

Pons
- Locus coeruleus (LC)

Fibers
- Frontal white matter (FW)
- Optic chiasm (oh)
- Capsula interna (CI)
- Corpus callosum (CC)

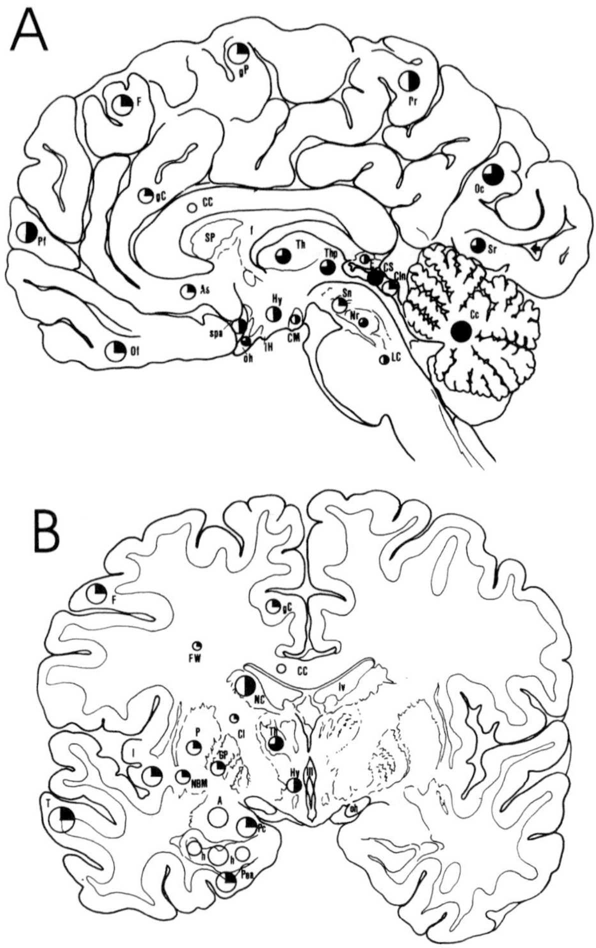

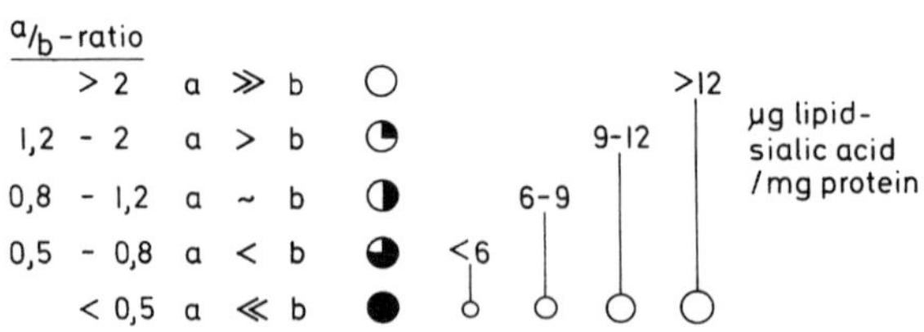

Fig. 6. Area-specific expression of gangliosides in the adult human brain demonstrated in a longitudinal (**A**) and a transversal (**B**) section. The total ganglioside-bound sialic acid per mg protein is indicated by the size, the a- to b-ganglioside ratios by the *black* portions of the cycles. Note that a-pathway gangliosides predominate in the hippocampus (*h*), amygdala (*A*), and corpus callosum (*CC*), whereas in the occipital neocortex (*OC*), cerebellum (*Cc*), thalamus (*Th*), and colliculi superiores (*CS*) b-pathway gangliosides are prevalent

GM1b and GD1 have been reported to be developmentally regulated (Hirabayashi et al. 1990). In the rat and mouse brain, O-acetylated forms of GD3 and GD2 were found to be transiently expressed by radial glial cells and migrating neurons (Mendez-Otero et al. 1988; Schlosshauer et al. 1988; Kotani et al. 1995; Santiago et al. 2001).

2.7 Aging and Neural Degeneration

Only a few groups have investigated profiles of brain gangliosides throughout aging, revealing partly different results. Whereas Segler-Stahl et al. (1983) reported a 50% decrease in total gangliosides of the whole human brain (based on fresh weight) occurring between ages 25 and 50, Svennerholm et al. (1989) found in the human frontal cortex no remarkable change during this period. Likewise, we were not able to detect dramatic changes in total ganglioside-content of the human frontal, occipital, and cerebellar cortex, and hippocampus between ages 20 and 50 (Kracun et al. 1992). Thereafter, up to age 90, however, a more or less pronounced decrease in total gangliosides was observed. With respect to individual ganglioside types, we found a substantial decrease in GD1a, whereas GD1b and GT1b remain rather constant. As a consequence of these slow changes, a shift of a prevalence of a-gangliosides (GM1, GD1a) to a predominance of b-gangliosides (GD1b, GT1b) occurs in the aging human brain (Fig. 2). A similar shift from a- to b-gangliosides was observed in the aging rat brain (Hilbig et al. 1984) and thus seems to be a general feature of the aging mammalian brain. Also, c-polysialogangliosides, which are abundant in the embryonic brain, are detectable in small amounts in human brain throughout the life span. Interestingly, a c-ganglioside-related epitope was immunohistochemically detected in fibrillary senile plaques and neurofibrillary tangles of brains with Alzheimer's disease by means of the mAbs A2B5 (Emory et al. 1987), M7103 and M6704 (Takahashi et al. 1991).

3 Possible Functional Roles of Gangliosides in the Developing Nervous System

Up to now, there is no clear-cut evidence for a defined functional role of the complex ganglio-series gangliosides in nervous tissue development. This critical statement is substantially supported by the report that mice with the disrupted GM2/GD2 synthase gene and lacking complex gangliosides did not show any obvious morphological (including myelination) defects in their nervous system or behavior, except for a slight reduction in the neural conduction velocity in the tibial nerve (Takamiya et al. 1996). However, these mutants expressed much higher levels of GM3 and GD3, and it is possible that these simpler gangliosides may somehow substitute for more complex gangliosides in whatever roles gangliosides play in normal brain development. On the other hand, the pronounced enrichment of gangliosides in neuronal membranes and an increasing body of functional evidence obtained from neuronal cell lines and nonneural cell systems suggest that these glycolipids may play important roles in the normal development of the central nervous system, too. In general, three modes of action of gangliosides affecting cell signalling, including the regulation of cell proliferation, differentiation and growth, have been proposed: (1) modulation of growth factor/hormone receptors and integrin receptors as well as cell/cell and cell/extracellular matrix interaction by gangliosides, (2) induction of signals together with transducer molecules by gangliosides organized in cell membrane microdomains, and (3) induction of signals by primary degradation products or derivatives of gangliosides (Sena 1993; Nagai and Tsuji 1994; Hakomori et al. 1998; Igarashi 1998; Spiegel et al. 1998; Yates and Rampersand 1998; Prinetti et al. 1999, 2000).

Much indirect evidence suggests that the above-proposed mode of ganglioside actions may also be important for normal nervous tissue development. Thus, numerous authors reported the effects of exogenous gangliosides on embryonic and regenerative neuronal growth (Ledeen et al. 1998). Concerning endogenous gangliosides, neuronal GD1a and GT1b were found to function as ligands for myelin-associated glycoprotein, promoting myelin stability and controlling nerve regeneration (Yang et al. 1996; Vyas and Schnaar 2001). Another specific function of ganglioside GT1b in neurons was claimed by Kotani et al. (2000), who identified p58, a brain-specific, sodium-dependent inorganic phosphate cotransporter and a GT1b-binding protein. In cell cultures of primary neurons and neuroblastoma cells, GM1 was found to be upregulated in the nuclear membrane following Ca^{2+}-elevating stimulation and facilitates the efflux of nuclear Ca^{2+} (Ledeen et al. 1998). Including data from genetic diseases – Tay-Sachs disease and related neuronal storage disorders – ganglioside GM2 is strongly suggested to be specifically involved in pyramidal neuron dendritogenesis (Walkley et al. 1995, 1998), although the mechanisms are still unknown. Likewise, a specific functional involvement of O-acetylated GD3 (and GD2) in neuronal migration in the mammalian brain is most likely. However, also in this case, the mechanisms are not yet clarified (Mendez-Otero and Santiago 2001).

Taken together, a still increasing amount of data clearly demonstrates striking and characteristic changes and a pronounced accretion of certain gangliosides to correlate with progress in nervous tissue development. The functional meaning of these cell surface-expressed glycosphingolipids for neurons and glial cells and their interactions, however, still remains to be elucidated in detail.

References

Asou H, Hirano S, Kyemura K (1989) Ganglioside composition of astrocytes. Cell Struct Funct 14:61–568

Bouvier JD, Seyfried TN (1989) Ganglioside composition of normal and mutant mouse embryos. J Neurochem 52:60–466

Cheresh DA, Pierschbacher MD, Herzig MA, Mujoo K (1986) Disialogangliosides GD2 and GD3 are involved in the attachment of human melanoma and neuroblastoma cells to extracellular matrix proteins. J Cell Biol 102:88–696

Chou KH, Nolan CE, Jungalwala FB (1982) Composition and metabolism of gangliosides in rat peripheral nervous system during development. J Neurochem 39:1547–1558

Cochran FB, Yu RK, Ledeen RW (1982) Myelin gangliosides in vertebrates. J Neurochem 39:773–779

Constantine-Paton M, Blum AS, Mendez-Otero R, Barnstable CJ (1986) A cell surface molecule distributed in a dorsoventral gradient in the perinatal rat retina. Nature 324:459–462

Daniotti JL, Landa CA, Gravotta D, Maccioni HJF (1990) GD3 ganglioside is prevalent in fully differentiated neurons from rat retina. J Neurosci Res 26:436–446

Daniotti JL, Landa CA, Rösner H, Maccioni HJF (1991) GD3 prevalence in adult rat retina correlates with the maintenance of a high GD3-/GD2-synthase activity ratio throughout development. J Neurochem 57:2054–2058

Daniotti JL, Landa CA, Rösner H, Maccioni HJF (1992) Adult rat retina interneurons synthesize GD3: GD3 expression by these cells is regulated by cell–cell interactions. J Neurochem 59:107–117

Derrington E, Borroni E (1990) The developmental expression of the cholinergic-specific antigen Chol-1 in the central and peripheral nervous system of the rat. Dev Brain Res 52:131–140

Dreyfus H, Urban PF, Edelharth S, Mandel P (1975) Developmental patterns of gangliosides and of phospholipids in chick retina and brain. J Neurochem 25:245–250

Dreyfus H, Louis JC, Harth S, Mandel P (1980) Gangliosides in cultured neurons. Neuroscience 6:1647–1655

Emory CR, Ala TA, Frey WH (1987) Ganglioside monoclonal antibody (A2B5) labels Alzheimer's neurofibrillary tangles. Neurology 37:768–772

Engel EL, Wood JG, Byrd FI (1979) Ganglioside patterns and cholera toxin–peroxidase labeling of aggregating cells from the chick optic tectum. J Neurobiol 10:429–440

Eto Y, Shinoda S (1982) Gangliosides and neutral glycosphingolipids in human brain tumors: specificity and their significance. In: Makita A (ed) New vistas in glycolipid research. Plenum, New York, pp 279–290

Felding-Habermann B, Jennemann R, Schmitt J, Wiegandt H (1986) Glycosphingolipid biosynthesis in early chick embryos. Eur J Biochem 166:651–658

Fishman PH, Brady RO (1976) Biosynthesis and function of gangliosides. Science 194:906–915

Fredman P, Dumanski J, Davidsson P, Svennerholm L, Collins VP (1990) Expression of the ganglioside GD3 in human meningiomas is associated with monosomy of chromosome 22. J Neurochem 55:1838–1841

Goldman JE, Hirano M, Yu RK, Seyfried TN (1984) GD3 ganglioside is a glycolipid characteristic of immature neuroectodermal cells. J Neurochem 7:179–192

Greis C, Rösner H (1990a) C-pathway polysialogangliosides in the nervous tissue of vertebrates, reacting with the monoclonal antibody Q211. Brain Res 571:105–110

Greis C, Rösner H (1990b) Migration and aggregation of embryonic chicken neurons in vitro: possible functional implication of polysialogangliosides. Dev Brain Res 57:223–234

Hakomori S (1981) Glycosphingolipids in cellular interaction, differentiation, and oncogenesis. Annu Rev Biochem 50:733–764

Hakomori S, Yamamura S, Handa K (1998) Signal transduction through glyco(sphingo)lipids. Introduction and recent studies on glyco(sphingo)lipid-enriched microdomains. Ann NY Acad Sci 845:1–10

Hamburger V, Hamilton HL (1951) A series of normal stages in the development of the chick embryo. J Morphol 50:49–92

Hilbig R, Rahmann H (1987) Phylogeny of vertebrate brain gangliosides. In: Rahmann H (ed) Gangliosides and modulation of neuronal functions. NATO ASI series 7. Springer, Berlin Heidelberg New York, pp 333–350

Hilbig R, Rösner H, Merz G, Segler-Stahl C, Rahmann H (1982) Developmental profiles in mouse and rat cerebral cortex. Roux's Arch Dev Biol 191:281–284

Hilbig R, Lauke G, Rahmann H (1984) Brain gangliosides during the life span (embryogenesis to senescence) of the rat. Dev Neurosci 6:260–270

Hirabayashi Y, Hirota M, Matsumoto M, Tanaka H, Obata K, Ando S (1988a) Developmental changes of C-series polysialogangliosides in chick brains revealed by mouse monoclonal antibodies M6704 and M7103 with different epitope specificities. J Biochem 104:937–979

Hirabayashi Y, Nakao T, Matsumoto M, Obata K, Ando S (1988b) Improved method for large-scale purification of brain gangliosides by Q-Sepharose column chromatography. Immunochemical detection of C-series polysialogangliosides in adult bovine brains. J Chromatogr 445:377–384

Hirabayashi Y, Hyogo A, Nakoa T, Tsuchiya S, Suzuki Y, Matsumoto M, Kon K, Ando S (1990) Isolation and characterization of extremely minor gangliosides, GM1b and GD1 alpha, in adult bovine brains as developmentally regulated antigens. J Biol Chem 265:8144–8151

Iber H, van Echten G, Klein D, Sandhoff K (1990) pH-dependent changes of ganglioside biosynthesis in neuronal cell culture. Eur J Cell Biol 52:236–240

Igarashi I (1998) Sphingosine-1-phosphate as an intercellular signaling molecule. Ann NY Acad Sci 845:19–31

Irwin LN, Michael DB, Irwin CC (1980) Ganglioside patterns of fetal rat and mouse brain. J Neurochem 34:1527–1530

Ishizuka J, Wiegandt H (1972) An isomer of trisialoganglioside and the structure of tetra- and pentasialogangliosides from fish brain. Biochem Biophys Acta 260:279–289

Kim SU, Morotto G, Yu RK (1986) Neuroimmunology of gangliosides in human neurons and glial cells in culture. J Neurochem 15:303–321

Klenk E (1942) †ber die Ganglioside, eine neue Gruppe von zuckerhaltigen Gehirnlipoiden. Hoppe-Seyler's Z Physiol Chem 273:76–86

Kojima N, Hakomori S (1991) Cell adhesion, spreading, and motility of GM3-expressing cells based on glycolipid-glycolipid interaction. J Biol Chem 266: 17552–17558

Kotani M, Kawashima I, Ozawa H, Terashima T, Tai T (1993) Differential distribution of major gangliosides in rat central nervous system detected by specific monoclonal antibodies. Glycobiol 3:137–146

Kotani M, Kawashima I, Ozawa H, Ogura K, Ishizuka I, Terashima T, Tai T (1994) Immunohistochemical localization of minor gangliosides in the rat central nervous system. Glycobiology 4:855–865

Kotani M, Terashima T, Tai T (1995) Developmental changes of ganglioside expressions in postnatal rat cerebellar cortex. Brain Res 700:40–58

Kotani M, Tajima Y, Shimoda Y, Irie A, Kubo H, Tai T (2000) Ganglioside GT1b in rat brain binds to p58, a brain-specific sodium-dependent inorganic phosphate cotransporter: expression cloning with a specific monoclonal antibody to ganglioside GT1b-binding protein. J Biochem (Tokyo) 127:13–22

Kracun I, Rösner H, Kostovic J, Rahmann H (1983) Areal and laminar distribution of gangliosides in the fetal human neopallium at 28 weeks of gestation. Roux's Arch Dev Biol 192:108–112

Kracun I, Rösner H, Cosovic C, Stavljenic A (1984) Topographical atlas of the gangliosides of the adult human brain. J Neurochem 43:979–989

Kracun I, Rösner H, Cosovic C (1986) Topographical distribution of the gangliosides in the developing and adult human brain. In: Tettamanti G, Ledeen RW, Sandhoff K, Nagai J, Toffano G (eds) Neuronal plasticity and gangliosides. Plenum, New York, pp 339–348

Kracun I, Rösner H, Drnovsek V, Heffer-Lauc M, Cosovic C, Lauc G. (1991) Human brain gangliosides in development, aging and disease. Int J Dev Biol 35:289–295

Kracun I, Rösner H, Drnovsek V, Vukelic Z, Cosovic C, Trbojevic-Cepe M, Kubat M (1992) Gangliosides in the human brain development and aging. Neurochem Int 20:421–431

Landa CH, Moskona AA (1985) Changes in ganglioside profile in chick embryo retina: studies on tissue and cell cultures. Int J Dev Neurosci 3:77–78

Landa CH, Panzetta P, Maccioni HJF (1984) Biosynthesis of gangliosides in cultured retina from chick embryos. Dev Brain Res 14:83–92

Ledeen RW (1978) Ganglioside structures and distribution: are they localized at the nerve ending? J Supramol Struct 8:1–17

Ledeen RW, Yu RK (1982) Gangliosides: structure, isolation, and analysis. In: Ginsburg V (ed) Methods in enzymology. Academic Press, New York, pp 139–191

Ledeen RW, Wu G, Lu ZH, Kozireski-Chuback D, Fang Y (1998) The role of GM1 and other gangliosides in neuronal differentiation. Overview and new finding. Ann NY Acad Sci 845:161–175

Letinic K, Heffer-Lauc M, Rösner H, Kostonic I (1998) C-pathway polysialogangliosides are transiently expressed in the human cerebrum during fetal development. Neuroscience 86:1–5

Levine SM, Seyfried TN, Yu RK. and Goldman JE, (1986) Immunocytochemical localization of GD3 ganglioside to astrocytes in murine cerebellar mutants. Brain Res 374: 260–269

Maccioni HJF, Panzetta P, Arrieta D, Caputto R (1984) Ganglioside glycosyltransferase activities in the cerebral hemispheres from developing rat embryos. Int J Dev Neurosci 2:13–19

Maccioni HJF, Landa CA, Panzetta P (1989) Developmental regulation of ganglioside biosynthesis. Studies in the chick embryo retina. Neurol Neurobiol 49:117–127

Mansson JE, Vanier MT, Svennerholm L (1978) Changes in the fatty acid and sphingosine composition of the major gangliosides of human brain with age. J Neurochem 30:273–275

Mendez-Otero R, Constantine-Paton M (1990) Granule cell induction of 9-O-acetyl gangliosides on cerebellar glia in microcultures. Dev Biol 138:400–409

Mendez-Otero R, Santiago M (2001) Functional role of a glycolipid in directional movements of neurons. An Acad Bras Cienc 73:221–229

Mendez-Otero R, Schlosshauer B, Barnstable CJ, Constantin-Paton M (1988) A developmentally regulated antigen associated with neural cell and process migration. J Neurosci 8:564–579

Merat A, Dickerson JWT (1973) The effect of development on the gangliosides of rat and pig brain. J Neurochem 20:873–880

Nagai Y, Tsuji S (1994) Significance of ganglioside-mediated glycosignal transduction in neuronal differentiation and development. Prog Brain Res 101:119–126

Nagai Y, Sanai Y, Nakaishi H (1987) Fundamentals of genetic control of gangliosides: the enigma of carbohydrate chain diversity in glycosphingolipids. In: Rahmann H (ed) Gangliosides and modulation of neuronal functions. NATO ASI Series 7. Springer, Berlin Heidelberg New York, pp 275–292

Nakaishi H, Sanai Y, Shiroki K, Nagai Y (1988) Analysis of cellular expression of gangliosides by gene transfection. I: GD3 expression in myc-transfected and transformed 3Y1 correlates with anchorage-independent growth activity. Biochem Biophys Res Commun 150:760–765

Ogawa-Goto K, Abe T (1998) Gangliosides and glycosphingolipids of peripheral nervous system myelins – a minireview. Neurochem Res 23:305–310

Panzetta P, Maccioni HJF, Caputto R (1980) Synthesis of retinal gangliosides during chick embryonic development. J Neurochem 35:1001–08

Panzetta PD, Gravotta D, Maccioni HJF (1987) Biosynthesis and expression of gangliosides during differentiation of chick embryo retina cells in vitro. J Neurochem 49:1763–1771

Pohlentz G, Klein D, Schwarzmann G, Schmitz D, Sandhoff K (1988) Both GA2, GM2, and GD2 synthases and GM1b, GD1a, and GT1b synthases are single enzymes in Golgi vesicles from rat liver. Proc Natl Acad Sci USA 85:7044–7048

Prinetti A, Iwabuchi K, Hakomori S (1999) Glycosphingolipid-enriched signaling domain in mouse neuroblastoma Neuro2a cells. Mechanism of ganglioside-dependent neuritogenesis. J Biol Chem 274:20916–20924

Prinetti A, Chigonora V, Tettamati G, Sonnino S (2000) Sphingolipid-enriched membrane domains from rat cerebellar granule cells differentiated in culture. A compositional study. J Biol Chem 275:11658–11665

Probstmeier R, Pesheva P (1999) Tenascin-C inhibits beta1 integrin-dependent cell adhesion and neurite outgrowth on fibronectin by a disialoganglioside-mediated signaling mechanism. Glycobiology 9:101–114

Pukel CS, Lloyd KO, Travassos LR, Dippold WG, Oettgen HF, Lloyd JO (1982) GD3, a prominent ganglioside of human melanoma. Detection and characterisation by mouse monoclonal antibody. J Exp Med 155:1133–1147

Rebhan M, Vacun G, Bayreuther K, Rösner H (1994) Altered ganglioside expression by SH-SY5Y cells upon retinoic acid-induced neuronal differentiation. Neuro Rep 5:941–944

Reynolds R, Wilkin GP (1988) Expression of GD3 ganglioside by developing rat cerebellar Purkinje cells in situ. J Neurosci 20:311–319

Richardson PJ, Walker JH, Jones RT, Whittaker VP (1982) Identification of a cholinergic-specific antigen Chol-1 as a ganglioside. J Neurochem 38:1605–1614

Robert J, Rebel G, Mandel P (1977) Glycosphingolipids from cultured astroblasts. J Lipid Res 18:517–522

Rohrer H, Henke-Fahle S, El-Sharkaway T, Lux HD, Thoenen HC (1985) Progenitor cells from embryonic chick dorsal root ganglia differentiate in vitro to neurons: biochemical and electrophysiological evidence. EMBO J 4:1709–1714

Rösner H (1975) Changes in the contents of gangliosides and glycoproteins and in the ganglioside pattern of the perinatal chicken brain. J Neurochem 24:815–816

Rösner H (1977) Sialoglycoproteins and acetylcholinesterase of the developing mouse brain. Roux's Arch Dev Biol 183:325–335

Rösner H. (1980) Gangliosides changes in the chicken optic lobes and cerebrum during embryonic development. Transient occurrence of "novel" multisialogangliosides. Roux's Arch Dev Biol 188:205–213

Rösner H (1981) Isolation and preliminary characterization of novel polysialogangliosides from embryonic chick brain. J Neurochem 37:993–997

Rösner H (1982) Ganglioside changes in the chicken optic lobes as biochemical indicators of brain development and maturation. Dev Brain Res 236:49–61

Rösner H, Al-Aqtum M, Henke-Fahle S (1985a) Developmental expression of GD3 and polysialogangliosides in embryonic chicken nervous tissue reacting with monoclonal antiganglioside antibodies. Dev Brain Res 18:85–95

Rösner H, Rahmann H, Reuter G, Schauer R, Katalinic JP, Egge H. (1985b) Mass spectrometric identification of the pentasialoganglioside GP1c of embryonic chicken brain. Biol Chem Hoppe-Seyler 366:1177–1181

Rösner H, Willibald CJ, Henke-Fahle S (1986) Expression of ganglioside-antigens during neuronal differentiation studied by use of monoclonal antibodies. In: Tettamanti G, Ledeen RW, Sandhoff K, Nagai J, Toffano G (eds) Neuronal plasticity and gangliosides. Plenum Press, New York, pp 330–338

Rösner H, Greis C, Henke-Fahle S (1988a) Developmental expression in embryonic rat and chicken brain of a polysialoganglioside-antigen reacting with the monoclonal antibody Q211. Dev Brain Res 42:161–171

Rösner H, Greis C, Willibald CJ, Henke-Fahle S (1988b) Developmental expression of gangliosides in nervous tissue. In: Ledeen R, Tettamanti G, Yu RK, Hogan E, Yates A (eds) New trends in ganglioside research. Springer, Berlin Heidelberg New York, pp 435–448

Rösner H, Al-Aqtum M, Rahmann H (1992) Gangliosides and neuronal differentiation. Neurochem Int 20:339–351

Sandhoff K, Christomanou H (1979) Biochemistry and genetics of gangliosidoses. Humangenetic 50:107–143

Sandhoff K, Schwarzmann G, Sarmientos F, Conzelmann E (1987) Fundamentals of ganglioside catabolism. In: Rahmann H (ed) Gangliosides and modulation of neuronal functions. NATO ASI series 7. Springer, Berlin Heidelberg New York, pp 231–250

Santiago MF, Berredo-Pinho M, Costa MR, Gandra M, Cavalcante LA, Mendez-Oetro R (2001) Expression and function of ganglioside 9-O-acetyl GD3 in postmitotic granule cell development. Mol Cell Neurosci 17:488–499

Sbaschnig-Agler M, Dreyfus H, Norton WT, Sensenbrenner M, Farooq M, Burne MC, Ledeen RW (1988) Gangliosides of cultured astroglia. Brain Res 461:98–106

Schaal H, Wille C, Wille W (1985) Changes of ganglioside pattern during cerebellar development of normal and staggerer mice. J Neurochem 45:544–551

Schlosshauer B, Blum AS, Mendez-Otero R, Barnstable CJ, Constantine-Paton M (1988) Developmental regulation of ganglioside antigens recognized by the JONES antibody. J Neurosci 8:580–592

Schnaar RL, Collins BE, Wright LP, Kiso M, Tropak MB, Roder JC, Crocker PR (1998) Myelin-associated glycoprotein binding to gangliosides. Structural specificity and functional implications. Ann NY Acad Sci 845:92–105

Segler-Stahl K, Webster JC, Brunngraber EG (1983) Changes in the concentration and composition of human brain gangliosides with aging Gerantology 29:161–168

Sena A (1993) Gangliosides in neurobiology. Acta Med Rep 6:341–346

Seybold V, Rahmann H (1985) Changes in developmental profiles of brain gangliosides during ontogeny of a teleost fish (*Sarotherodon mossambicus*, Cichlidae). Roux's Arch Dev Biol 194:166–172

Seyfried TN (1987) Ganglioside abnormalities associated with failed neural differentiation in a T-locus mutant mouse embryo. Dev Biol 123:286–291

Seyfried TN, Yu RK (1985) Ganglioside GD3: structure, cellular distribution, and possible function. Mol Cell Biochem 68:3–10

Sonnino S, Bassi R, Chignoro V, Tettamanti G (1990) Further studies on the changes of chicken brain gangliosides during prenatal and postnatal life. J Neurochem 54:1653–1660

Sparrow JR, Barnstable CJ (1988) A gradient molecule in developing rat retina: expression of 9-O-acetyl GD3 in relation to cell type, developmental age, and GD3 ganglioside. J Neurosci 21:398–409

Spiegel S, Cuvillier O, Edsallki C, Kohama T, Menzelke R, Olah Z, Olivera A, Pirianov G, Thomas DM, Tu Z, VanBrocklyn JR (1998) Sphingosine-1-phosphate in cell growth and cell death. Ann NY Acad Sci 845:11–18

Suzuki K (1965) The pattern of mammalian brain gangliosides. II. Evaluation of the extraction procedures, postmortem changes and the effect of formalin preservation. J Neurochem 12:969–979

Svennerholm L (1963) Chromatographic separation of human brain gangliosides. J Neurochem 10: 613-623

Svennerholm L, Boström K, Fredman P, Mansson JE, Rosengren B, Rynmark BM (1989) Human brain gangliosides: developmental changes from early fetal stage to advanced stage. Biochim Biophys Acta 1005:109–117

Takahashi H, Hirokowa K, Ando S, Obata K (1991) Immunohistological study on brains of Alzheimer's disease using antibodies to fetal antigens, C-series gangliosides and microtubule-associated protein 5. Acta Neuropathol (Berl) 81:626–631

Takamiya K, Yamamoto A, Furukawa K, Yamashiro S, Shin M, Okoda M, Fukumoto S, Haraguchi M, Takeda N, Fujimura K, Sakae M, Kishikawa M, Shiku H, Furukawa K, Aizawa S (1996) Mice with disrupted GM2/GD2 synthase gene lack complex gangliosides, but exhibit only subtle defects in their nervous system. Proc Natl Acad Sci USA 93:10662–10667

Tettamanti G, Ghidoni R, Trinchera M (1987) Fundamentals of brain ganglioside biosynthesis. In: Rahmann H (ed) Gangliosides and modulation of neuronal functions. NATO ASI Series 7. Springer, Berlin Heidelberg New York, pp 191–204

Thangnipon W, Balazs R (1992) Developmental changes in gangliosides in cultured cerebellar granule neurons. Neurochem Res 17:45–59

Thierfelder S, Pini S, Harrison F, Wiegandt H (1992) Immunohistochemical localisation of monoclonal antibody R 24-recognized ganglioside Glac2 in early chick embryos. Differentiation 49:7–15

Van Echten G, Sandhoff K (1989) Modulation of ganglioside biosynthesis in primary cultured neurons. J Neurochem 52:207–214

Van Echten G, Iber H, Stotz H, Takatsuki A, Sandhoff K (1990) Uncoupling of ganglioside biosynthesis by Brefeldin A. Eur. J Cell Biol 51:135–139

Vanier MT, Holm M, Mansson JE, Svennerholm L (1973) The distribution of lipids in the human nervous system. V. Gangliosides and allied neutral glycolipids of infant brain. J Neurochem 21:1375–1384

Vyas AA, Schnaar RL (2001) Brain gangliosides: functional ligands for myelin stability and the control of nerve regeneration. Biochimie 83:677–682

Walkley SU, Siegel DA, Dobrenis K (1995) GM2 ganglioside and pyramidal neuron dendritogenesis. Neurochem Res 20:1287–1299

Walkley SU, Siegel DA, Dobrenis K, Zervas M (1998) GM2 ganglioside as a regulator of pyramidal neuron dendritogenesis. Ann NY Acad Sci 845:188–199

Wiegandt H (1982) The gangliosides. Adv Neurochem 4:149–223

Wiegandt H (1995) The chemical constitution of gangliosides of the vertebrate nervous system. Behav Brain Res 66:85–97

Yang LJ-S, Zeller CB, Shaper NL, Kiso M, Hasegawa A, Shapiros RE, Schnaar RL (1996) Gangliosides are neuronal ligands for myelin-associated glycoprotein. Proc Natl Acad Sci USA 93:814–818

Yates AJ, Rampersand A (1998) Sphingolipids as receptor modulators. An overview. Ann NY Acad Sci 845:57–71

Yavin E, Yavin Z (1979) Gangliosides profiles during neural tissues development. Acquisition in the prenatal rat brain and cerebral cell cultures. Dev Neurosci 2:25–37

Yu RK (1994) Development regulation of ganglioside metabolism Prog Brain Res 101:31–44

Yu RK, Ando S (1980) Structures of some new complex gangliosides of fish brain. Adv Exp Med Biol 125:35–45

Yu RK, Macula LJ, Taki T, Weinfeld HM, Yu FS (1988) Developmental changes in ganglioside composition and synthesis in embryonic rat brain. J Neurochem 50:1825–1829

Zervas M, Walkley S (1999) Ferret pyramidal cell dendritogenesis: changes in morphology and ganglioside expression during cortical development. J Comp Neurol 413:429–448

Zhang G, Kurono S, Fujita SC, Furuya S, Hirabayashi Y (1997) Developmentally regulated O-acetylated sialoglycans in the central nervous system revealed by a new monoclonal antibody 493D4 recognizing a wide range of O-acetylated glycoconjugates. Glycoconj J 14:847–857

Mannose-Binding Lectins in Cerebrum Development

J.-P. ZANETTA

1 Introduction

The nervous system is so complex that it is not ordinarily considered as an interesting model for studying cellular mechanisms. However, its morphology (at least in some well-defined areas) provides essential information because cells are polarised: cell body, axons and dendrites can be easily identified in vivo, an essential property when studies are concerned with the intracellular traffic of molecules and with the biochemical steps involved in cell adhesion and cell recognition mechanisms. The development of the central nervous system presents features that are phylogenetically preserved. Although most studies were performed on a special portion of the central nervous system (CNS), the cerebellum (Fig. 1), because its organisation is extremely simple, reproducible from one area to the other and its development is relatively synchronous compared with other brain areas, the mechanisms deduced from this simple part of the CNS can be extrapolated to other brain areas. The formation of the specialised contacts between neurones at the synapses is only one aspect of the complexity of the CNS development. In fact, the formation of the CNS can be reduced to sequential steps: neuronal cell proliferation, neuronal cell migration and axonal growth, synaptogenesis and glial wrapping (myelination and astrocytic insulation). In the cerebellum, after the initial phase of neuronal cell proliferation, the cell bodies of immature neurones undergo migration on pre-existing astrocytic fibres to reach their adult location. In the cerebellum (Altman 1972a–c; Palay and Chan-Palay 1974), precursors of the small neurones that are called "granule cells" proliferate in the "external granular layer" [or "external germinative layer"(egl)] close to the pia mater, then migrate to the adult position in the "internal granular layer"

Laboratoire de Glycobiologie Structurale et Fonctionnelle, CNRS UMR 8576, Batiment C9, 59655 Villeneuve d'Ascq Cédex, France

Progress in Molecular and Subcellular Biology, Vol. 32
Ivica Kostović (Ed.)

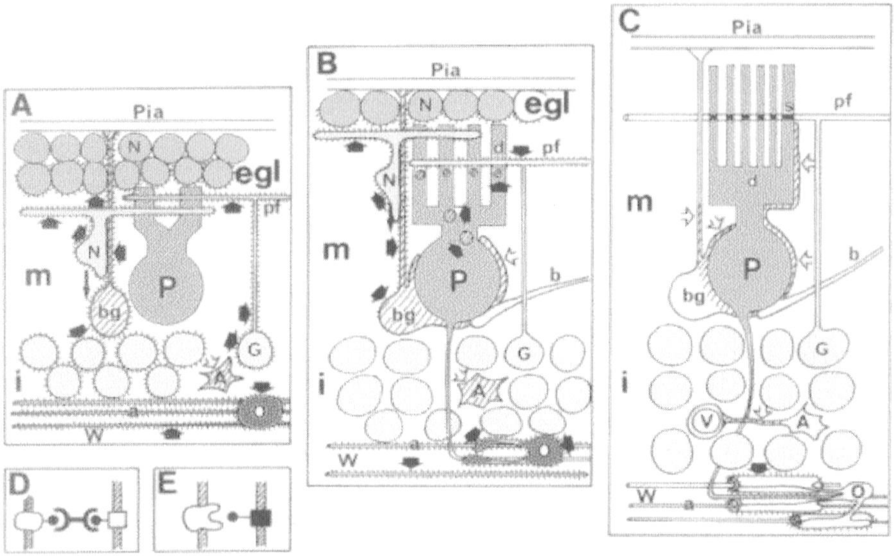

Fig. 1A–C. Schematic drawing of the cerebellar development in rat. **A** In the second postnatal week, the precursors of the granule cells (*N*) proliferate in the external germinative layer (*egl*), accumulate in this area (up to ten rows of cells), then migrate (*N*) through the molecular layer (*m*) along the Bergmann fibres making bridges between the pia mater (*Pia*) and the cell body of Golgi epithelioid cells (*bg*) localised below the Purkinje cell body (*P*), until the internal granular layer (*i*). During the migration, these neuroblasts begin to differentiate and grow axons parallel to the pia mater and perpendicular to the Purkinje cell dendritic tree, the parallel fibres (*pf*). **B** Once the granule cell bodies of the granule cells arrive in the internal granular layer, the parallel fibres make synapses with the spiny branchlets of the dendritic tree of the Purkinje cell (each parallel fibre makes about 400 synapses with the dendrites of different Purkinje cells). **C** Situation in the adult cerebellum. *O* Oligodendrocyte, *A* astrocyte, *b* axon of basket cells. **D** Scheme of the function of lectin CSL as a bridging molecule between the membrane of migrating neuroblasts and the Bergmann fibres. **E** Scheme of the interaction of R1 at the surface of Purkinje cell dendrites and its ligands at the surface of parallel fibres

(i). This migration occurs along processes (Bergmann fibres) of radial glial cells (the Golgi epithelioid cells) having a cell body close to the Purkinje cell body and processes forming *pseudopodia* on the pia mater. During this mechanism, called "contact guidance of neurone migration" (Rakic 1971; Sidman and Rakic 1973), neurones make intimate contact with the Bergman fibres, pushing a protoplasmic process along the astrocytic fibre, the nuclei migrating through this process. At the same time, the axon of the granule cell divides into two branches parallel to the pia mater, perpendicular to the Purkinje cell dendritic tree in the molecular layer, called the "parallel fibres". Once the cell body arrives in the internal granular layer, the parallel fibres make synapses with the dendritic spines of all neurones found in the molecular layer. This stage of synaptogenesis is followed by a glial wrapping of neuronal elements (wrapping of large neurones and clusters of granule cells by astrocytic walls, wrapping of large axons by myelin (axons of basket cells and Purkinje cells and incoming axons, the mossy fibres and climbing fibres in the white matter and in the internal granular layer i).

The hypothesis of the involvement of the interactions between endogenous lectins and glycoconjugates in adhesion and recognition in the CNS was hypothesised a long time ago, but the evidence for the presence of "endogenous lectins" in higher organisms than plants was contested. The discovery of a β-galactoside-binding lectin from electric organ tissue of *Electrophorus electricus* (Teichberg et al. 1975) was an important indication that these carbohydrate-binding proteins, supposed for a long time to be specific for lower organisms or plants (RŸdiger 1998), could be present in animals. Today, hundreds of lectins have been isolated from mammals that were demonstrated to play essential roles in intracellular traffic, cell adhesion, cell recognition mechanisms or in the regulation of cell proliferation in most tissues so far analysed, including the nervous and immune systems. Disease-associated alterations of the structure of glycoconjugates, readily determined by elaborated analysis procedures including lectins as tools (Brink et al. 1998; Danguy et al. 1998; Geyer and Geyer 1998) and widely documented (Brockhausen 1998; Hakomori 1998) can thus be tentatively set into relation to a cognitive protein-carbohydrate interplay. Besides the actual classical C-type lectins, galectins and Siglec-type lectins (Drickamer 1988, 1997; Gabius 1997), several studies emphasised that molecules of the superfamily of cytokines may have lectin activities. In the latter case (Zanetta et al. 1996, 1998a, b; Cebo et al. 2001, 2002), the small cytokines have both a receptor-binding domain and a carbohydrate-recognition domain and behave as bi-functional molecules, the two domains being necessary for their biological function. This chapter will summarise the function of mannose-binding lectins in the nervous tissue, with special focus on the developmental events in the CNS and, especially, in the cerebellum.

2
Evidence for the Presence of Central Nervous System Mannose-Binding Lectins

2.1
Ontogenetic Variations of Glycoconjugates During Central Nervous System Development

In a systematic study of the glycoprotein-bound monosaccharides during the development of the rat cerebellum, it was observed that mannose of the membrane-bound glycoproteins showed an important accumulation during the second and third post-natal week compared with the other monosaccharides, suggesting that mannose-rich glycoproteins were accumulated during this period (Zanetta et al. 1978). Isolation of these putative glycoproteins by affinity chromatography on Concanavalin A (ConA) in the presence of SDS verified this accumulation, and indicated that the accumulation was only concerned with a few glycoprotein subunits (Zanetta et al. 1978). These transiently expressed glycoproteins (termed C1 glycoproteins) were not identified at this time, but corresponded to known antigens: two of these subunits, initially termed B4 and B4 (Zanetta et al. 1978) corresponded to the rat equivalent of the human CD24, a GPI-anchored glycoprotein considered to be a good marker of human B cells. One of these glycoproteins, initially termed B2, corresponded to the murine Thy-1 antigen, a GPI-anchored glycoprotein considered to be a marker of murine T cells. The fact that these glycoproteins were GPI-anchored explained their quasi insolubility in neutral detergents, a property that appeared thereafter to be extremely important for the isolation of the cerebellar mannose-binding lectins.

The first question was asked about the mechanism of the disappearance of these glycoproteins, which constituted, at the 13th post-natal day, about 35% of the total proteins of the rat cerebellum. One possibility was that the mannose-rich glycans of these glycoproteins were completed by more external monosaccharides, a point that was theoretically possible at this time because of the absence of clear knowledge on the biosynthesis of N-glycans. The other possibility was that these glycoproteins were actually destroyed. This was actually the case, since labelling experiments indicated that the lifetime of the glycans of the glycoproteins synthesised between the 10th and the 16th post-natal day was about 78 h, whereas the lifetime of the glycoproteins synthesised after this period was greater than 1 month. This indicated that these glycoproteins were massively synthesised and, after 3 days, were massively degraded (Reeber et al. 1980).

Based on the histochemical staining of these transiently expressed glycoproteins (using ConA both at the level of the optical and electron microscope), it was evident that these glycoproteins were concentrated at the surface of axons, and especially at the surface of parallel fibres, the axons of granule cells, the quantitatively major neuronal cell type in the cerebellum. In fact, these glycoproteins were massively accumulated at the membrane of all parts of the granule cells (cell body and growing axons) as soon as they left the proliferative zone in the external granular layer. Since these glycoproteins were found at the surface of moving structures (migrating neuronal cell bodies and growing axons), it was suggested that these compounds could play a role as anti-adhesive molecules impeding interactions with the surrounding extracellular matrix and/or cells. This was at least unlikely for the migrating neurones. Indeed, morphological studies indicated that migrating granule cells make intimate contact with the Bergmann fibres, the processes of the Golgi epithelioid cells (the radial astrocytes localised below the Purkinje cell body) which make *pseudopodia* on the cerebellar pia mater. These morphological studies rather suggested a role of these glycoproteins in the contact guidance of neuronal migration.

3 Degradation of the Accumulated Axonal Mannose-Rich Glycoproteins in the Target Neurones

The next question was asked to determine where the C1 glycoproteins of the membrane of parallel fibres were degraded. Indeed, there was no evidence in the literature for the presence of glycolytic and proteolytic enzymes in the extracellular compartment and the parallel fibres did not contain lysosomes or other structures likely to contain such degradative enzymes. Thus, the idea emerged that these glycoproteins could be degraded into the cells with which these axons would form synapses. Since the disappearance of these glycoproteins occurred at a period corresponding to the first stages of synaptogenesis, the concept that the disappearance of the glycoprotein was the elimination of the recognition signals once the recognition had occurred was likely. In fact, several morphological studies reported that the early stage of synaptogenesis in all parts of the cerebrum was characterised by a specific abundance of phagocytotic processes of the membranes of axons by their target cells (Altman 1971; Eckenhoff and Pysh 1979). This process, not observed in the adult in all parts of the CNS, could reappear during a neo-synaptogenesis subsequent to lesions of the afferent fibres (Eckenhoff and Pysh 1979). We, therefore, hypothesised that the C1 glycoproteins of the parallel fibres could be degraded into their target cells.

In order to test this hypothesis, we treated pups with chloroquine, a substance known to elevate the pH of the endosomal and lysosomal compartment. The result of this treatment was a very intense accumulation of material in non-functional lysosomes of Purkinje cells (Dontenwill et al. 1983). This material was stained intensely with ConA, indicating the abundance of mannose-rich glycoproteins. This material did not issue from Purkinje cells, since we did not observe any staining for the PSG antigen (Reeber et al. 1981; Langley et al. 1982). This mannose-rich glycoprotein was found to be absolutely specific for Purkinje cells within the whole cerebrum (with the exception of one cell with a Purkinje morphology localised in the zone of proliferation of these cells during the pre-natal period). The PSG antigen is distributed on all RER, Golgi membranes and sublemmal cisternae and never found at the Purkinje cell surface. It was more recently identified as the rat equivalent of the human CD3 (Zanetta et al. 1996), a marker of human T cells.

Therefore, these studies indicated that the degradation of the axonal C1 glycoproteins occurred in their target cells. This suggested that a specific α-mannosidase could be present in these cells, which could be over-expressed at this period. This was actually the case, both for the enzymatic activity and for the immuno-histochemical localisation of this antigen after isolation and production of polyclonal antibodies (Zanetta et al. 1982, 1983).

In the experiments of the chloroquine treatments of rats, we tested the behaviour of several other surface glycoproteins of the surface of parallel fibres and, especially, 5′-nucleotidase, another ConA-binding glycoprotein. In fact (Dontenwill et al. 1983), this glycoprotein was not accumulated in the target cells. Since the internalisation mechanism (in fact phagocytosis of portions of the membranes of parallel fibres by their target cells) was specific for these glycoproteins, it was suggested that this internalisation was a receptor-mediated internalisation and, consequently, that mannose-binding lectins were present.

4 Evidence for the Presence of Mannose-Binding Lectins in the Cerebrum

Several methods were used in order to obtain evidence for the presence of mannose-binding lectins. One of them was the use of neoglycoproteins (i.e. synthetic polymers [poly-lysine, poly-ornithine or proteins (bovine serum albumin)] substituted by a high number (15–25) monosaccharides and labelled either by fluorescent dyes (for fluorescence microscopy) or with biotin (Monsigny et al. 1983; SΣve et al. 1986; Gabius et al. 1989) for light and subsequent electron microscope examinations). Such techniques were found to be very efficient for the detection of putative lectin activities in cells or tissue sections (Bardosi et al. 1988, 1990a, b; Gabius and Bardosi 1990; Gabius et al. 1988, 1989, 1991, 1993; Debagge et al. 1990). In some cases, the comparison of the localisation of already identified lectins was performed between such techniques and immuno-cytochemical techniques with a relatively good correlation (Kuchler et al. 1990, 1992a. However, the possibility remained that endogenous lectins were inactivated by aldehyde treatments necessary to fix the tissue for morphological studies. The possibility also existed that the majority of the carbohydrate-recognition domains of the lectins was involved in the binding of the high-affinity endogenous oligosaccharide ligands and, consequently, was no more accessible to the glycoconjugate tools. In fact, this technique was not used in our initial experiments designed to detect the presence of endogenous mannose-binding lectins, but was used later on to compare this methodology with the data of the immuno-histochemical localisation of actually identified mannose-binding lectins. As reported elsewhere (Kuchler et al. 1990, 1992a), a good correlation was observed between the neoglycoprotein technique and the immuno-cytochemical techniques.

For the initial detection of the mannose-binding lectins we made the hypothesis that during tissue homogenisation the endogenous mannose-binding lectins will all bind to their endogenous ligands, i.e. the C1 glycoproteins. Since these glycoproteins were highly insoluble in neutral detergents, the complexes between mannose-binding lectins and the C1 glycoproteins should also be insoluble and recovered as an insoluble residue after extraction of most proteins and glycoproteins not involved in this interaction using neutral detergents. In contrast, the complex between mannose-binding lectins and their ligands should be dissociated using high mannose concentrations. A refinement of this technique by using this type of protocol first without detergent allowed us to identify the "cerebellar soluble lectin" CSL (Zanetta et al. 1987a), whereas a subsequent protocol in the presence of neutral detergent allowed the purification of the "membrane-bound" lectin R1 (Dontenwill et al. 1985; Zanetta et al. 1985).

5 The Cerebellar Soluble Lectins

The CSLs constitute a family of immunologically related molecules endowed with identical carbohydrate-binding properties. These molecules are oligomers of high Mr (more than 10^6 K; Zanetta et al. 1987a), formed of the assembly of identical subunits. The latter (Mr 31.5, 33 and 43 K) show a cellular specificity. The CSLs recognise specifically $Man_6GlcNAc_2Asn$, but have no affinity for the same glycan lacking the first GlcNAc residue (Marschal et al. 1989). The reason for this apparent discrepancy is that in $Man_6GlcNAc_2Asn$, the Man(α1--2)Man(α1–3) branch interacts with the di-N-acetyl-chitobiose moiety of the core as determined by NMR on the CD2 molecule (Wyss et al. 1995). Therefore, CSL recognises a specific conformation of this glycan, which is neither found in other oligomannosidic glycoasparagines nor on oligomannosides possessing a single GlcNAc residue in the reducing position. Due to their oligomeric structures, the CSL molecules show very strong agglutinating activities. For example, fixed human red blood cells covalently coated with RNAse B glycoasparagines (containing essentially $Man_5GlcNAc_2Asn$ and $Man_6GlcNAc_2Asn$) are agglutinated using concentrations as low as 0.15 μg/ml CSL (instead of 4 μg/ml for the tetravalent ConA; unpubl. data). Consequently, the CSLs can be extremely efficient adhesion molecules for cells having specific oligomannosidic ligands, when it is externalised by cells at different periods of ontogenetic processes.

5.1 Lectin Cerebellar Soluble Lectins and Contact Guidance of Neurone Migration

Several studies performed on cultured cerebellar explants provided clues for the understanding of the molecular mechanisms involved in the contact guidance of neurones. The technique used consisted in the ^{3}H-thymidine labelling of proliferating neurones (around the 10th post-natal day in the rat) and to localise the neurones by autoradiography after 3 days on sections of the explants. Using this system, it was demonstrated that the migration was not inhibited by addition of the different glycosaminoglycans, nor by neuraminidase treatment, indicating that GAGs, gangliosides or sialylated N- and O-glycans are not involved (Lindner et al. 1986). This migration was not significantly inhibited by Fab fragments of antibodies against the HNK-1 epitope (glucuronic acid 3-sulfate) or against the poly Neu5Ac(α2–8) found on the embryonic N-CAM. In contrast, this migration was inhibited partially using anti-L2/NgCAM Fab fragments (Lindner et al. 1983) and totally using the anti-L3 antibody (Fahrig et al. 1990). This epitope recognised by this antibody has been defined as the oligomannosidic oligosaccharide $Man_6GlcNAc$, indicating an essential role of this glycan.

In fact, the migration was completely inhibited using small amounts of Fab fragments against CSL (Lehmann et al. 1990). The CSL is produced by astrocytes and externalised. It recognises several glycoprotein ligands at the astrocyte surface and several ligands at the surface of migrating neurones, the major ones being Thy-1 and the rat equivalent of the human CD24 (Kuchler et al. 1989a; Kay et al. 1991). The later is a 31 amino acid polypeptide endowed with a glycosylphosphatidylinositol anchor and three N-glycans, and represents the major molecular entity of the 13-day-old rat cerebellum (Zanetta et al. 1978). It is transiently expressed with a life-time of about 3 days. The migration of cerebellar neurones is also 90% inhibited using small amounts of the Fab fragments directed against this molecule (Lehmann et al. 1990). In conclusion, these data indicate that the contact between migrating neurones and radial astrocyte fibres involves formation of bridges between the two structures through an extracellular soluble polyvalent lectin and its oligomannosidic ($Man_6GlcNAc_2Asn$) membrane-bound ligands at the surface of the two cells in contact. Because of the postulated bridging role attributed to CSL, it was expected that the Fab fragments of the anti-CSL antibodies could totally inhibit the neurone migration. This was less evident for the Fab fragments of the anti-CD24 antibodies, because other ligands (Thy-1) were present. In fact, ultrastructural pictures obtained using the ConA-peroxidase technique or the anti-CD24 antibody indicated that these ligands are organised into clusters (Kuchler et al. 1989b), which could correspond to the attachment point of lectin CSL, the latter showing a similar ultrastructural cell surface distribution. Because of these clusters of GPI-anchored glycoproteins, it can be understood that an antibody against one of these constituents can have a very important steric hindrance of the interaction of CSL with its oligosaccharide ligands in the complex.

As also observed for the process of myelination, the fact that the CSL ligands are GPI-anchored may be of importance because of the absence of interactions of these ligands with the neuronal cytoskeleton. Consequently, the CSL-induced contact can ensure a strong and specific adhesion process of the neuronal surface with astrocytes, but a free mobility of molecules at the surface of migrating neurones.

5.2 Lectin Cerebellar Soluble Lectin and Myelination Processes

Lectin CSL (45-K-form) produced and externalised by myelinating cells, oligodendrocytes and myelinating Schwann cells plays different roles in the process of myelination. Interestingly, the non-myelinating Schwann cells present in the PNS do not express CSL (Kuchler et al. 1992b).

The first function of CSL is that of a mitogen for myelinating cells. It acts as an autocrine/paracrine mitogen for these cells, the externalised CSL inducing the clustering of its surface ligands (the major one is the "early form" of the myelin-associated glycoprotein MAG; Fig. 2) that provokes specific phosphorylations on Tyr and Ser/Thr residues (Badache et al. 1995). These could be induced by addition of exogenous CSL and normal and CSL-induced proliferation/phosphorylations can be inhibited by anti-CSL Fab fragment. The role of CSL in this mechanism can be understood as an agent clustering all surface molecular complexes containing the CSL ligands. The other is the mitogenic activity of these cells induced by the contact with axonal membranes. Although it was assumed that the mitogen was a soluble heparan sulfate proteoglycan of the axonal membrane (because of its solubilisation using high heparin concentrations; Ratner et al. 1988), the mitogenic activity can be solubilised from axonal membranes using lower concentrations of $Man_6GlcNAc_2Asn$. The solubilised material contained in fact, only CSL immunoreactive material (Badache et al. 1995). CSL-depleted axonal membrane did not show mitogenic activities, but this mitogenic activity of CSL-depleted axonal membranes could be restored by the addition of exogenous CSL to axonal membranes. Such CSL bound to axonal membrane ligands shows a higher mitogenic activity than the soluble CSL, probably because of the pre-clustering of CSL

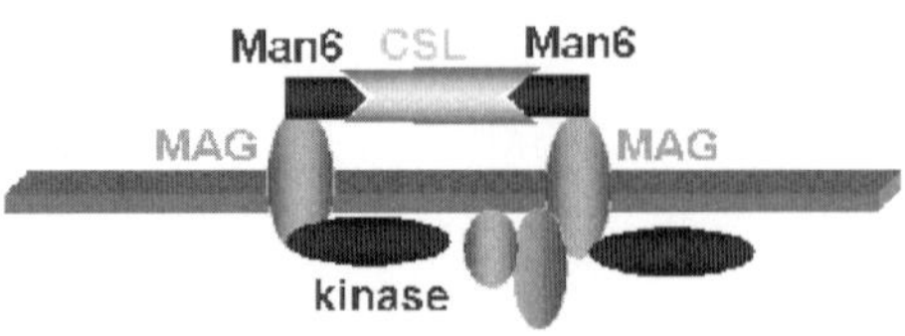

Fig. 2. Schematic drawing of the signal transducing role of CSL on myelinating cells. The polyvalent lectin CSL (in fact, 40 identical subunits instead of 2 represented here) externalised by myelinating cells produces a clustering of its glycoprotein ligands at the surface of myelinating cells. This clustering induced specific phosphorylations at the origin of the proliferative signal. The major glycoprotein ligand of CSL is the "early" form of MAG, possessing the $Man_6GlcNAc_2Asn$ (Man6) ligand of CSL

bound to axonal membranes that favour the clustering of the CSL ligands at the surface of myelinating cells (Badache et al. 1995).

The second function, related to the latter, is concerned with the initial contact between axons and myelinating cells (Fig. 3). During the first stage of myelination, myelinating cell processes establish contact for a relatively long distance with axons. In these areas, ultrastructural studies (Kuchler et al. 1989a, b) showed a co-localisation of extracellular CSL and its major axonal ligand, the rat equivalent of CD24. Since ligands of CSL ("early form" of MAG, L1/Ng-CAM) have been localised at the surface of myelinating cells in the same area (Martini and Schachner 1986; Ishiguro et al. 1991), it is suggested that CSL makes bridges between ligands at the surface of these two structures, as a first contact for the initiation of myelination (Kuchler et al. 1989a). After this initial stage, the constituents of this interaction disappear in large part, but the same constituents remain in adult myelin, restricted to the single area of contact between axons and myelinating cells, i.e. the lateral loops of the Ranvier's node (Kuchler et al. 1989a). These ultrastructurally special junctional complexes (freeze fracture experiments showed a ball-bearing structure on the axonal surface) are the unique adult brain regions where CD24 can be detected, co-localised with CSL immunoreactive material. The fact that axonal ligands of CSL (Thy-1 and CD24) are GPI-anchored in the axonal membrane (without relationship with the axonal cytoskeleton) suggests that these junctions can actually function as ball-bearing, a situation necessary to explain the absence of twisting of the axons during myelination (Kuchler et al. 1989b).

The third function is the compaction of myelin. Indeed, low amounts of CSL are present in compact myelin both in the central and peripheral nervous tissue. The identified ligand of adult compact myelin in the

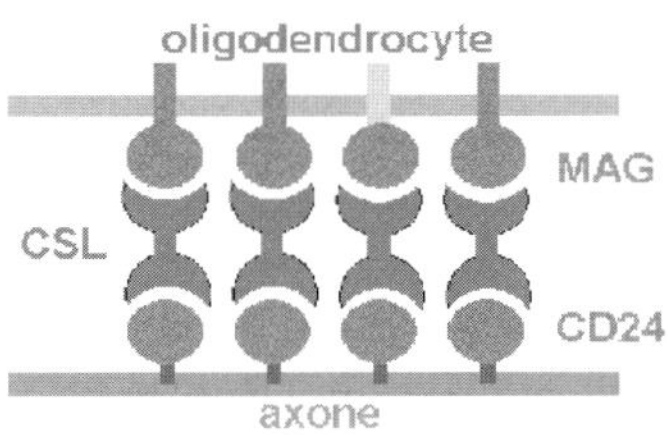

Fig. 3. Scheme of the mechanism of action of CSL as a bridging molecule between axons and myelinating cells. CSL produced and externalised by myelinating cells makes bridges between its ligands at the axonal surface (the major one being CD24) and at the surface of the myelinating cell (the major one being the "early" form of MAG). This mechanism occurring for long distances in the first period of myelination, is restricted to the junctional area of the lateral loops of the Ranvier's nodes in the adult

PNS is the P0 glycoprotein (but, apparently, only a small proportion of P0 molecules; Kuchler et al. 1989b; Badache et al. 1992). In the CNS, minor low M_r glycoproteins have been identified as CSL ligands. The demyelinating activity of anti-CSL Fab fragments on myelin produced by cultured oligodendrocytes suggests that CSL is actually involved in myelin compaction (Kuchler et al. 1988). These functions of CSL in the myelination processes suggest that it could be an important immunological target in demyelinating diseases, especially multiple sclerosis (MS; Zanetta 1996). In fact, most of the patients with MS (more than 95%) have anti-CSL antibodies in their CSF, whereas less than 15% of patients with other neurological diseases do not (Zanetta et al. 1990, 1994). Interestingly, the anti-CSL antibodies found in the CSF of MS patients are, at least in large majority, inhibitors of the lectin activity of CSL and, consequently, are potentially demyelinating (unpubl. data). In contrast, the majority of the anti-CSL antibodies found in the CSF of the false positive group were not inhibitory of the lectin activity of CSL. Recent studies suggest that these inhibitory antibodies can be, at least in part, responsible for the demyelination observed in MS. Indeed, the intraventricular injection in adult Wistar rats (insensitive to EAE) of a monoclonal anti-CSL antibody inhibiting the lectin activity of CSL is followed by their specific invasion of the area of the *corpus callosum* and a subsequent demyelination with ultrastructural and cellular event observed for the periventricular demyelination in MS (unpubl. data). The reasons why the anti-CSL antibodies penetrate the parenchyma are two-fold: (1) CSL is involved in the formation of a tight junction between ependymal cells lining ventricles and anti-CSL antibodies are able to disrupt these junctions (Perraud et al. 1988); (2) these antibodies are generally blocked by the underlying astrocytic layer, except in one area where this layer is absent: the area of the *corpus callosum*.

6
The Lectin R1

The calcium-independent lectin R1 was initially isolated from the developing rat cerebellum as a mannose-binding lectin (Zanetta et al. 1985). This lectin shows a duality of high specificity for oligomannosides of the $Man_5GlcNAc$ and $Man_8GlcNAc$ type (Marschal et al. 1989), the reasons for this duality are not yet determined. Throughout the neural cells, R1 is specifically localised in neurones (Zanetta et al. 1985). It shows a transient over-expression in large cerebellar neurones, especially in the Purkinje cells between the 10th and 21st post-natal day in the rat. During this period (Fig. 4), the intracellularly accumulated R1

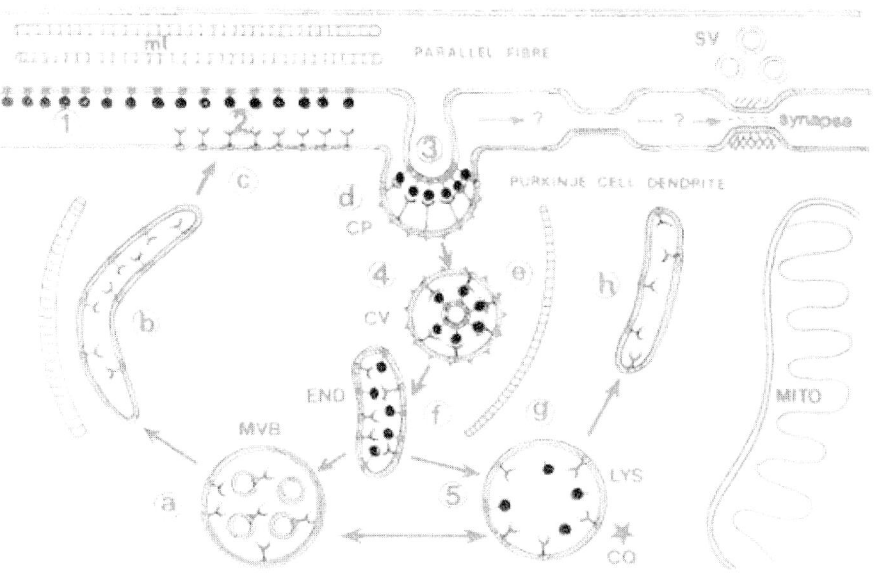

Fig. 4. Proposed scheme for the role of lectin R1 and of the C1 glycoproteins in the mechanism of neuronal recognition preceding synaptogenesis. *1* The C1 glycoproteins (CD24 and Thy-1; *red circles*) are accumulated at the surface of the parallel fibres. *2* Lectin R1 is synthesised in Purkinje cells, transported to the spiny branchlets of some portion of the dendritic tree of Purkinje cells and externalised. *3* After recognition of the C1 glycoproteins by R1, the complex provokes a coated invagination (coated pits; *cp*) of the membrane of Purkinje cell together with a portion of the membrane of parallel fibres. *4* This portion is internalised into the Purkinje cell dendrite through double-walled coated vesicles. *5* A compartmentalisation of C1 glycoproteins and R1 occurs in the endosomal/multivesicular body compartment. This process is blocked by chloroquine (*CQ*), R1 and the C1 glycoproteins being recovered in the non-functional lysosomes. In the normal situation, part of R1 is segregated in MVB (*a*), transported intracellularly to other regions of the dendritic tree of Purkinje cells (*b*) and externalised (*c*)

is transported to certain portions of the dendritic tree and is externalised (Dontenwill et al. 1985). It specifically recognises glycans of a few glycoproteins on the surface of incoming parallel fibres, the major ones being the Thy-1 and CD24 antigens (these two molecules represent more than 25% of the total cerebellar protein entities at this period). This contact is followed by the specific receptor R1-mediated internalisation of these glycoproteins into Purkinje cells, where glycoproteins are degraded (Dontenwill et al. 1983, 1985; Zanetta et al. 1991), whereas the lectin seems to be recycled to other parts of the Purkinje cell dendritic tree (Fig. 4). The internalisation of these glycoproteins occurs though “double walled coated vesicles”, structures which were shown to be specifically abundant during the early stage of synaptogenesis (Altman 1971; Eckenhoff and Pysh 1979). Ultrastructural studies suggest that these vesicles fuse with the endosomal compartment in complex structures called multivesicular bodies (Zanetta et al. 1987b), with a targeting of glycoproteins to lysosomes. An artificial increase in the pH of the endosomal and lysosomal compartments using chloroquine (Dontenwill et al. 1983) induces, at this period, a dramatic accumulation of R1 and its ligands in non-functional lysosomes. Experiments designed for impairing the normal cerebellar development [(absence of thyroid hormone, treatment with tunicamycin (Zanetta et al. 1987b, 1992)] suggest that R1 and its glycoprotein ligands participate in the recognition mechanism associated with the first step of synaptogenesis. In the case of tunicamycin, the late degeneration of neurones not having succeeded making synapse with their targets resembles the "en cascade" degeneration observed in the *staggerer* mutant (Sotelo and Changeux 1974). The hypothesis of the role of lectin R1 in neuronal recognition has been substantiated by the finding that the two complementary molecules (lectin R1 at the surface of the dendrites of Purkinje cells and the P31 glycoprotein at the surface of parallel fibres) are re-expressed in adult rat cerebella after 24–48 h during the repair of sections of parallel fibres (Lehmann et al. 1993). Lectin R1-mediated internalisation of the same axonal glycoproteins occurs also through "double walled coated vesicles" as previously observed by Chen and Hillman (1982) as a first step of synapse repair in the adult animal. This mechanism may provide an explanation for the increased number of "double walled coated vesicles" observed at the ultrastructural level during the period preceding synaptogenesis in various parts of the CNS (Altman 1971; Eckenhoff and Pysh 1979). However, the stabilisation of the specific contact in the synapse is certainly supported by another mechanism, since the lectin R1 is not present in mature synapses (absence in synaptic densities and synaptic junctions).

The transient expression of R1 at the surface of neurones is not restricted to the cerebellum, but occurs, with different timing, in all part of the CNS. A specific study in the PNS (Thomas et al. 1994) showed that this lectin is expressed in muscle (also target cells of neurones) and that it is involved in myoblast fusion. However, later on, it is also likely to be involved in the formation of the neuromuscular junction, based on experiments of sections of the sciatic nerve in the adult rat. In these conditions, the early forms of R1, totally absent in the adult, reappear. In these lesion experiments (unpubl. data), one ligand of R1 (the rat equivalent of CD24) is rapidly re-expressed, but first by the sensitive neurone of the reflex arc, then by motoneurones.

The structure of this lectin, initially defined as a "membrane-bound" lectin is still puzzling and its membrane-bound nature remains still operational. Indeed, the 67-K-form obtained by SDS-PAGE was split into two electrophoretic subunits of 45 and 23 K after immuno-purification on a column of a monoclonal antibody raised against the 67 form. This suggested a priori a proteolytic cleavage in acidic conditions used for the antigen elution. However, sequencing of the 45-K-subunit showed an identity with rat actin. Therefore, the proteolysis hypothesis was rejected and replaced (since the eluted material still preserved its lectin activity) by the hypothesis that the R1 lectin was made of a stable complex between actin and a calcium-independent mannose-binding protein. Such a molecule was actually found in slime mold and termed as comitin (Jung et al. 1996), a molecule considered as involved (in poorly polarised cells) in the intracellular traffic. The sequence of this molecule was extremely similar to that of the mannose-binding lectin of *Galanthus nivalis*, a calcium-independent mannose-binding lectin. In fact, the rat brain R1 molecule can be isolated by immuno-affinity on a column of anti-*Galanthus nivalis* antibodies, the purified material reacting with anti-R1 antibodies and having, in cell cultures, the same localisation of authentic R1 (unpubl. data).

6.1 Interleukin-2 and Oligodendrocyte Proliferation

Interleukin-2 was shown to produce the proliferation of oligodendrocytes (Jiang and Lu 1998), these cells expressing the beta chain of the interleukin-2 receptor (IL-2Rβ). Although it is not known if IL-2 is produced by the nervous tissue and if circulating IL-2 can cross the blood brain barrier to stimulate oligodendrocyte proliferation, molecules are present on oligodendrocytes at their endogenous period of proliferation (beginning of myelination) that can explain such proliferative activity. IL-2 is a calcium-independent lectin endowed with a high affinity for $Man_{5-6}GlcNAc$ oligomannosides (Zanetta et al. 1996). Thus IL-2 appears as a calcium-independent lectin a little bit less specific than CSL (which recognises only $Man_6GlcNAc_2$ oligomannosides), but all ligands of CSL are ligands of IL-2. As previously described (Badache et al. 1995), the surface of proliferating oligodendrocytes possesses glycoproteins having $Man_6GlcNAc_2Asn$ identified as ligands of the endogenous lectin CSL. These glycans, transiently expressed during this period, are found as the N-glycans of the "early form of MAG" and on other high M_r glycoproteins. The action of IL-2 as a mitogen for oligodendrocytes can be explained by a mechanism similar to that proposed for the activation of

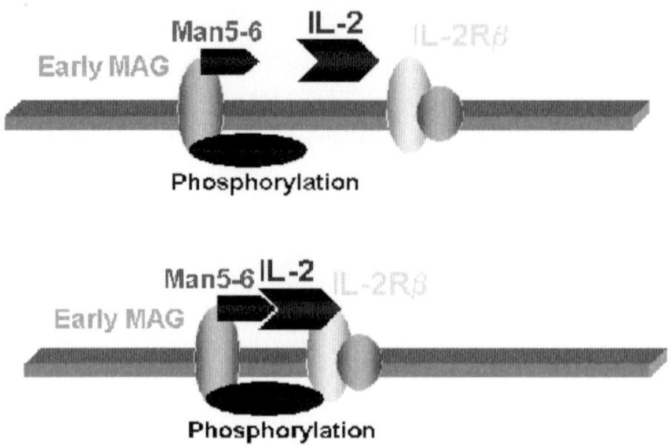

Fig. 5. Schematic drawing of the mitogenic action of interleukin-2 for myelinating cells. The bi-functional IL-2 bond to its receptor is able to associate (through its CRD) the later with its ligand, the major one being the "early" form of MAG

human T cells (Fig. 5): the bi-functional IL-2 binds to its β-receptor associates extracellularly, the latter with surface complexes containing glycoprotein ligands of IL-2, especially the early form of MAG. This association induced the intracellular contact between IL-2Rβ and kinases initially associated with the surface complexes containing the IL-2 ligands (Zanetta et al. 1996). In resting human T cells, the function of the lectin activity of IL-2 bound to its β-receptor is to associate the IL-2Rb with the T-cell receptor complex, IL-2 recognising an N-glycosylated form of CD3. Very recently (Fukushima et al. 2001), it has been shown in a cell line constitutively expressing both IL-2Rβ and IL-2Rα that IL-2Rα is not a receptor of IL-2, but a glycoprotein ligand of IL-2, possessing oligomannosidic N-glycans with 5 and 6 mannose residues. This specific interaction is responsible for the generation of a specific signal responsible for the IL-2 proliferation of these cells. Although fine experiments on myelinating cells are not yet accessible, it can be predicted that the proliferative activity of IL-2 on these cells is due to a specific clustering between the IL-2Rβ of these cells and the major ligand of IL-2, i.e. the early form of MAG is shown, using biotinylated CSL, to be necessarily a ligand of IL-2.

7 Conclusions and Perspective

Although a large variation of glycoconjugates was observed during ontogenetic development of the brain, their functional role remains undetermined. Several lectins (detected by immuno-histochemical techniques or using neoglycoproteins) showed specific and transient localisations. Specific roles were suggested for a few number of them. For example, galectins seem to play a role in the guidance of axonal growth (Mahanthappa et al. 1994). Mannose-binding lectins, however, play crucial roles in contact guidance of neuronal migration, interneuronal recognition, myelination and tightening of the ependymal cell barrier. This seems a little bit strange because oligomannosides are quite common structures in glycoprotein N-glycans. In fact, those oligomannosides showing considerable ontogenetic variations are restricted to certain forms of oligomannosides (e.g. the CSL ligands, $Man_6GlcNAc_2Asn$). Despite the large distribution of these compounds, their role in a tissue may be very specific since it is mediated through the expression of their lectin partner at the cell surface. In other words, cell recognition mechanisms could result from a synchronous expression of the two types of complementary molecules specifically at the site of the interaction. Furthermore, the glycoconjugate partners of lectins may concern several molecular entities bearing the same glycan and the same anchoring mode. This redundancy of the glycoprotein lectin ligand (CD24/Thy-1) for migration and synaptogenesis, early-MAG/L1-NgCAM for myelination may be necessary to preserve a fundamental function that would be easily suppressed by a single gene mutation or knock-out if the system used a single ligand.

References

Altman J (1971) Coated vesicles and synaptogenesis. A developmental study in the cerebellar cortex of the rat. Brain Res 30:311–322

Altman J (1972a) Postnatal development of the cerebellar cortex in the rat. I. The external germinal layer and the transitional molecular layer. J Comp Neurol 145:353–398

Altman J (1972b) Postnatal development of the cerebellar cortex in rat. II. Phases in the maturation of Purkinje cells and of the molecular layer. J Comp Neurol 145:399–464

Altman J (1972c) Postnatal development of the cerebellar cortex in the rat. III. Maturation of the components of the granular layer. J Comp Neurol 145:465–514

Badache A, Burger D, Villarroya H, Robert Y, Kuchler S, Steck AS, Zanetta J-P (1992) Carbohydrate moieties of MAG, P0 and other myelin glycoproteins potentially involved in cell adhesion. Dev Neurosci 14:342–350

Badache A, Lehmann S, Kuchler-Bopp S, Hand N, Zanetta J-P (1995) An endogenous lectin and its glycoprotein ligands are triggering basal and axon-induced Schwann cell proliferation. Glycobiology 5:371–383

Bardosi A, Dimitri T, Gabius H-J (1988) Endogenous carbohydrate binding proteins in oligodendrogliomas. A histochemical study. Acta Neuropathol (Berl) 446:1–7

Bardosi A, Bardosi L, Hendrys M, Wosgien B, Gabius H-J (1990a) Spatial differences of endogenous lectin expression within the cellular organisation of the human heart.

A glycohistochemical immunohistochemical and glycobiochemical study. Am J Anat 188:409–418
Bardosi A, Bardosi L, Lindenblatt R, Gabius H-J (1990b) Detection and mapping of endogenous receptors for carrier-immobilized constituents of glycoconjugates (lectins) by labelled (neo)glycoproteins and by affinity chromatography in human adult mesencephalon, pons, medulla oblongata and cerebellum. Histochemistry 94:285–291
Brink U, Korabiowska M, Bosbach R, Gabius H-J (1998) Detection of inflammation- and neoplasia-associated alterations in human large intestine by plant/invertebrate lectins, galectin-1 and neoglycoproteins. Acta Anat 161:219–233
Brockhausen I (1998) Glycoproteins and their relationship to disease. Acta Anat 161:371–374
Cebo C, Dambrouck T, Maes E, Laden C, Strecker G, Michalski J-C, Zanetta J-P (2001) Recombinant human interleukins, namely IL-1alpha, IL-1beta, IL-4, IL-6 and IL-7, show different and specific calcium-independent carbohydrate-binding properties. J Biol Chem 276:5685–5691
Cebo C, Durier V, Lagant P, Maes E, Florea D, Lefebvre T, Strecker G, Vergoten G, Zanetta J-P (2002) Function and molecular modeling of the interaction between human interleukin 6 with its HNK-1 oligosaccharide ligands. J Biol Chem 277:12246–12252
Chen S, Hillman E (1982) Plasticity of the parallel fiber-Purkinje cell synapse by spine takover and new synapse formation in the adult rat. Brain Res 240:205–220
Danguy A, Decaestecker C, Genten F, Salmon I, Kiss R (1998) Applications of lectins and neoglycoconjugates in histology and pathology. Acta Anat 161:206–218
Debbage P, Marguth F, Gabius HJ 81990) Glycohistochemical visualisation of mannose-binding lectins in the microvasculatures of brain and brain tumours. Act Histochem 40:113–116
Dontenwill M, Devilliers G, Langley OK, Roussel G, Hubert P, Reeber A, Vincendon G, Zanetta J-P (1983) Arguments in favour of endocytosis of glycoprotein components of the membrane of parallel fibers by Purkinje cells during the development of the rat cerebellum. Dev Brain Res 10:287–299
Dontenwill M, Roussel G, Zanetta J-P (1985) Immunohistochemical localization of a lectin like molecule “R1” during the postnatal development of the rat cerebellum. Dev Brain Res 17:245–252
Drickamer K (1988) Two distinct classes of carbohydrate-recognition domains in animal lectins. J Biol Chem 263:9557–9560
Drickamer K (1997) Making a fitting choice: common aspects of sugar-binding sites in plant and animal lectins. Structure 5:465–468
Eckenhoff MF, Pysh JJ (1979) Double-walled coated vesicle formation: evidence for massive and transient conjugate internalization of plasma membranes during cerebellar development. J Neurocytol 8:623–638
Fahrig T, Schmitz B, Weber D, Kuchererehret A, Faissner A, Schachner M (1990) Two monoclonal antibodies recognizing carbohydrate epitopes on neural adhesion molecules interfere with cell interactions. Eur J Neurosci 2:153–161
Fruttiger M, Montag D, Schachner M, Martini R (1995) Crucial role for the myelin-associated glycoprotein in the maintenance of axon-myelin integrity. Eur J Neurosci 7:511–515
Fukushima K, Hara-Kuge S, Ideo H, Yamashita K (2001) Carbohydrate recognition site of interleukin-2 in relation to cell proliferation. J Biol Chem 276:7351–7351
Gabius H-J (1997) Animal lectins. Eur J Biochem 243:543–576

Gabius H-J, Bardosi A (1990) Regional differences in the distribution of endogenous receptors for carbohydrate constituents of cellular glycoconjugates especially lectins in cortex hippocampus basal ganglia and thalamus of adult human brain. Histochemistry 93:581–592

Gabius H-J, Kohnke B, Hellmann T, Dimitri T, Bardosi A (1988) Comparative histochemical and biochemical analysis of endogenous receptors for glycoproteins in human and pig peripheral nerve. J Neurochem 51:756–763

Gabius H-J, Wosgien B, Hendrys M, Bardosi A (1991) Lectin localization in human nerve by biochemically defined lectin-binding glycoproteins neoglycoprotein and lectin-specific antibody. Histochemistry 95:269–277

Gabius H-J, Gabius S, Zemlyanukhina TV, Bovin NV, Brinck J, Danguy A, Joshi SS, Kayser K, Schottelius J, Tietze LF, Vidal-Vanaclocha F, Zanetta J-P (1993) Reverse lectin histochemistry: Design and application of glycoligands for detection of cell and tissue lectins. Histol Histopathol 8:369–383

Gabius S, Hellmann KP, Hellmann T, Brinck J, Gabius H-J (1989) Neoglycoenzymes. A versatile tool for lectin detection in solid-phase assays and glycohistochemistry. Anal Biochem 182:447–451

Geyer H, Geyer R (1998) Strategies for glycoconjugate analysis. Acta Anat 161:18–35

Hakomori S (1998) Cancer-associated glycosphingolipid antigens: their structure, organization and function. Acta Anat 161:79–90

Ishiguro H, Sato S, Fujita N, Inuzuka T, Nakano R, Miyatake T (1991) Immunohistochemical localization of myelin-associated glycoprotein isoforms during the development in the mouse brain. Brain Res 563:288–292

Jiang CL, Lu CL (1998) Interleukin-2 and its effects in the central nervous system. Biol Signals Recept 7:148–156

Jung E, Fucini P, Stewart M, Noegel AA, Schleicher M (1996) Linking microfilaments to intracellular membranes: the actin-binding and vesicle-associated protein comitin exhibits a mannose-specific lectin activity. EMBO J 15:1238–1246

Kay R, Rosten PM, Humphries RK (1991) CD24 a signal transducer modulating B-cell activation responses is a very short peptide with a glycosyl phosphatidylinositol membrane anchor. J Immunol 147:1412–1416

Kuchler S, Fressinaud C, Sarlième LL, Vincendon G, Zanetta J-P (1988) Cerebellar soluble lectin is responsible for cell adhesion and participates in myelin compaction in cultured rat oligodendrocytes. Dev Neurosci 10:199–212

Kuchler S, Rougon G, Marschal P, Lehmann S, Reeber A, Vincendon G, Zanetta J-P (1989a) Localization of a transiently expressed glycoprotein in developing cerebellum delineating its possible ontogenetic roles. Neuroscience 33:111–124

Kuchler S, Herbein G, Sarlième LL, Vincendon G, Zanetta J-P (1989b) An endogenous lectin CSL interacts with major glycoprotein components in peripheral nervous system myelin. Cell Mol Biol 35:581–596

Kuchler S, Zanetta J-P, Vincendon G, Gabius H-J (1990) Detection of binding sites for biotinylated neoglycoproteins and heparin (endogenous lectins) during cerebellar ontogenesis in the rat. Eur J Cell Biol 52:87–97

Kuchler S, Zanetta J-P, Vincendon G, Gabius H-J (1992a) Detection of binding sites for biotinylated neoglycoproteins and heparin (endogenous lectins) during cerebellar ontogenesis in the rat: an ultrastructural study. Eur J Cell Biol 59:373–381

Kuchler S, Lehmann S, Vincendon G, Zanetta J-P (1992b) Endogenous lectin CSL involved in myelination is absent from non-myelinating Schwann cells. J Neurochem 58:1768–1772

Lai C, Brow MA, Nave K-A, Noronha AB, Quarles RH, Bloom FE, Milner RJ, Sutcliffe JG (1987) Two forms of 1B236/myelin-associated glycoprotein, a cell adhesion molecule for postnatal development, are produced by alternative splicing. Proc Natl Acad Sci USA 84:4337–4341

Langley OK, Reeber A, Vincendon G, Zanetta J-P (1982) Fine structural localization of a new Purkinje cell specific glycoprotein subunit: an immunoelectron microscopical study. J Comp Neurol 208:335–344

Lehmann S, Kuchler S, Theveniau M, Vincendon G, Zanetta J-P (1990) An endogenous lectin and one of its neuronal glycoprotein ligands are involved in contact guidance of neuron migration. Proc Natl Acad Sci USA 87:6455–6459

Lehmann S, Kuchler S, Gobaille S, Marschal P, Badache A, Reeber A, Vincendon G, Zanetta J-P (1993) Lesion-induced reexpression of neonatal recognition molecules in adult rat cerebellum. Brain Res Bull 30:515–521

Lindner J, Rathjen FG, Schachner M (1983) L1 mono- and polyclonal antibodies modify cell migration in early postnatal mouse cerebellum. Nature 305:427–430

Lindner J, Guenther J, Nick H, Zinser G, Antonicek H, Schachner M, Monard D (1986) Modulation of cell migration by a glia-derived protein. Proc Natl Acad Sci USA 83:4568–4571

Mahanthappa NK, Cooper DNW, Barondes SH, Schwarting GA (1994) Rat olfactory neurons can utilize the endogenous lectin, L-14, in a novel adhesion mechanism. Development 120:1373–1384

Marschal P, Reeber A, Neeser J-R, Vincendon G, Zanetta J-P (1989) Carbohydrate and glycoprotein specificity of two endogenous cerebellar lectins. Biochimie 71:645–653

Martini R, Schachner M (1986) Immunoelectron microscopic localization of neural cell adhesion molecules (L1, N-CAM, and MAG) and their shared carbohydrate epitope and myelin basic protein in developing sciatic nerve. J Cell Biol 103:2439–2448

Monsigny M, Kieda C, Roche A-C (1983) Membrane glycoproteins glycolipids and membrane lectins as recognition signals in normal and malignant cells. Biol Cell 47:95–110

Palay SL, Chan-Palay VC (1974) Cerebellar cortex. Cytology and organization. Springer, Berlin Heidelberg New York, pp 1–336

Perraud F, Kuchler S, Gobaille S, Labourdette G, Vincendon G, Zanetta J-P (1988) Endogenous lectin CSL is present on the membrane of cilia of rat brain ependymal cells. J Neurocytol 17:745–751

Puche AC, Poirier F, Hair M, Bartlett PF, Key B (1996) Role of galectin-1 in the developing mouse olfactory system. Dev Biol 179:274–287

Rakic P (1971) Neuron glia relationship during granule cell migration in developing cerebellar cortex. A Golgi and electron microscopy study in *Macacus rhesus*. J Comp Neurol 141:283–312

Ratner N, Hong D, Lieberman MA, Bunge RP, Glaser L (1988) The neuronal cell-surface molecule mitogenic for Schwann cells is a heparin-binding protein. Proc Natl Acad Sci USA 85:6992–6996

Reeber A, Vincendon G, Zanetta J-P (1980) Transient Concanavalin A-binding glycoproteins of the parallel fibers of the developing rat cerebellum: evidence for the destruction of their glycans. J Neurochem 35:1273–1277

Reeber A, Vincendon G, Zanetta J-P (1981) Isolation and immunohistochemical localization of a Purkinje cell-specific glycoprotein subunit from rat cerebellum. Brain Res 229:53–65

Rüdiger H (1998) Plant lectins – more than just tools for glycoscientists. Occurrence, structure and possible functions of plant lectins. Acta Anat 161:130–152

Sève A-P, Hubert J, Bouvier D, Bourgeois C, Midoux P, Roche A-C, Monsigny M (1986) Analysis of sugar binding sites in mammalian cell nuclei by quantitative flow microfluorimetry. Proc Natl Acad Sci USA 83:5997–6001

Sherblom AP, Sathyamoorthy N, Decker JM, Muchmore A (1989) IL-2 a lectin with specificity for high mannose glycopeptides. J Immunol 143:939–944

Sidman RL, Rakic P (1973) Neuron migration with special reference to developing human brain: a review. Brain Res 62:1–35

Sotelo C, Changeux J-P (1974) Trans-synaptic degeneration "en cascade" in the cerebellar cortex of staggerer mutant mouse. Brain Res 67:519–526

Teichberg VI, Silman I, Beitsch D, Resheff D (1975) A β-galactoside binding protein from electric organ tissue of *Electrophorus electricus*. Proc Natl Acad Sci USA 72:1383–1387

Thomas D, Lehmann S, Kuchler-Bopp S, Marschal P, Zanetta J-P (1994) Differential expression of an endogenous mannose-binding protein R1 during muscle development and regeneration delineating its role in myoblast fusion. Glycobiology 4:23–38

Villalobo A, Gabius H-J (1998) Signalling pathways for transduction of the initial message of the glycocode into cellular response. Acta Anat 161:110–129

Wyss DF, Choi JS, Li J, Knoppers MH, Willis KJ, Arulanandam ARN, Smolyar A, Reinherz EL, Wagner G (1995) Conformation and function of the N-linked glycan in the adhesion domain of human CD2. Science 269:1273–1278

Zanetta J-P (1996) Glycoproteins and lectins in immune demyelinating human diseases. In: Montreuil J, Schachter H, Vliegenthart JFG (eds) Glycoproteins and disease, vol 2. Elsevier, Amsterdam, pp 395–404

Zanetta J-P, Roussel G, Ghandour MS, Vincendon G, Gombos G (1978) Postnatal development of rat cerebellum: massive and transient accumulation of Concanavalin A binding glycoproteins in parallel fiber axolemma. Brain Res 142:301–319

Zanetta J-P, Meyer A, Dontenwill M, Basset P, Vincendon G (1982) Purification and properties of α-D-mannosidase from adult rat brain and interaction with its antibodies. J Neurochem 39:1601–1606

Zanetta J-P, Roussel G, Dontenwill M, Vincendon G (1983) Immunohistochemical localization of α-D-mannosidase during the cerebellar development of the rat. J Neurochem 40:202–208

Zanetta J-P, Dontenwill M, Meyer A, Roussel G (1985) Isolation and immunohistochemical localization of a lectin-like molecule from the rat cerebellum. Dev Brain Res 17:233–243

Zanetta J-P, Meyer A, Kuchler S, Vincendon G (1987a) Isolation and immunochemical study of a soluble cerebellar lectin delineating its structure and function. J Neurochem 49:1250–1257

Zanetta J-P, Dontenwill M, Reeber A, Vincendon G (1987b) Expression of recognition molecules in the cerebellum of young and adult rats. NATO ASI Ser H 2:92–104

Zanetta J-P, Warter J-M, Kuchler S, Marschal P, Rumbach L, Lehmann S, Tranchant C, Reeber A, Vincendon G (1990) Antibodies to cerebellar soluble lectin CSL in multiple sclerosis. Lancet 335:1482–1484

Zanetta J-P, Kuchler S, Lehmann S, Badache A, Maschke S, Marschal P, Dufourcq P, Vincendon G (1991) Cerebellar lectins. Int Rev Cytol 135:129–154

Zanetta J-P, Kuchler S, Lehmann S, Badache A, Maschke S, Thomas D, Dufourcq P, Vincendon G (1992) Glycoproteins and lectins in cell adhesion processes. Histochem J 24:791–804

Zanetta J-P, Tranchant C, Kuchler-Bopp S, Lehmann S, Warter J-M (1994) Presence of anti-CSL antibodies in the cerebrospinal fluid of patients. A sensitive and specific test in the diagnosis of multiple sclerosis. J Neuroimmunol 52:175–182

Zanetta J-P, Alonso C, Michalski J-C (1996) Interleukin 2 is a lectin that associates its receptor with the T cell receptor complex. Biochem J 318:49–53

Zanetta J-P, Bonaly R, Maschke S, Strecker G, Michalski J-C (1998) Hypothesis: immunodeficiencies in α-mannosidosis, mycosis, aids and cancer: a common mechanism of inhibition of the function of interleukin 2 by oligomannosides. Glycobiology 8(3):V–IX

Zanetta J-P, Bonaly R, Maschke S, Strecker G, Michalski J-C (1998) Differential binding of lectins IL-2 and CSL to *Candida albicans* and cancer cells. Glycobiology 8:221–225

Functional Role of Gangliosides in Neuronal Motility

R. Mendez-Otero, L.A. Cavalcante

1 Introduction

Cell migration occurs often during mammalian brain development and persists in a few regions in the adult brain. Defective migratory behavior of neurons is thought to be the underlying cause of several congenital cortical disorders. Thus, an understanding of the dynamics and molecular mechanisms of neuronal cell migration could expand our knowledge of normal development of the central nervous system (CNS) as well as help in deciphering the pathogenesis of various developmental disorders (Aicardi 1994; Raymond et al. 1994; Uher and Golden 2000; Walsh and Goffinet 2000).

In the developing CNS, most neuronal precursors are generated in specialized germinal zones adjacent to the ventricle, termed the ventricular zone. A wave of secondary neurogenesis occurs late in development, often in the perinatal period, and produces huge numbers of interneurons. Three primary sites of secondary neurogenesis have been described: the external granular cell layer (EGL) of the cerebellum, the subgranular layer of the dentate gyrus, and the subventricular zone (SVZ) in the telencephalon. Both the dentate gyrus and the SVZ maintain a proliferative population of stem cells throughout life and give origin to neurons destined for the hippocampal formation and olfactory bulb, respectively (Altman 1969a, b; Luskin 1993; Lois et al. 1996; Gage et al. 1998). Once generated, neurons migrate from their birthplace in the proliferative layers to their ultimate locations in the mature brain, utilizing distinct migration strategies to reach their final destination.

Directional movements occur often at all stages of morphogenesis of the CNS. One of the first descriptions of directional movements,

Instituto de Biofisica Carlos Chagas Filho, Universidade Federal do Rio de Janeiro, CCS, 21941–590 Rio de Janeiro, Brazil

Progress in Molecular and Subcellular Biology, Vol. 32
Ivica Kostović (Ed.)

obtained from static images, was that of the interkinetic nuclear migration of the neuroepithelium provided by Sauer (1936) and confirmed with the aid of ^{3}H-thymidine autoradiography almost a quarter of a century later (Sauer and Walker 1959). Although confined within a narrow extent of the tissue, this oscillatory movement and its dependence on microtubules (Messier 1978) may provide helpful insights on the mechanisms of a certain step of neuroepithelial cell/neuronal migration: the movement of the nucleus – nucleokinesis (Morris et al. 1998) – into a previously formed leading process.

In addition to the interkinetic nuclear migration of the pseudostratified neuroepithelium, extensive cell migration occurs in the developing and adult CNS and ensures that postmitotic immature neurons generated in the primary or secondary germinative zones acquire their proper positions in the mature brain. First, after terminal mitosis, immature neurons migrate towards the pial surface until they reach the various structural primordia where subsequent maturation and final differentiation occurs. This is the first mode of migration – radial migration – involving movement of the neuron orthogonally to the pial surface, with most of the neurons moving associated to a special glia, the radial glia. Guidance along radial glia is a common mechanism for the differentiation of projection neurons and establishment of a blueprint of laminated structures, such as the cerebral cortex and the cerebellum (for review, see Pearlman et al. 1998; Hatten 1999; Parnavelas 2000). This migration process is a complex developmental program involving a series of steps, including cell-cell recognition, cell-cell adhesion, cell motility, and finally detachment from the glial fibers once neurons reach their destination (Rakic et al. 1994; Nadarajah et al. 2001). The other general migration mode – tangential migration – involves the displacement of neuronal progenitors or postmitotic cells parallel to the pial surface. This displacement is a prominent feature in the migration of neuronal progenitors from the rhombic lip that are destined to brainstem nuclei or to the subpial aspect of the cerebellar cortex (Sidman and Rakic 1982; Rakic 1985; Bourrat and Sotelo 1988; de Diego et al. 2002), in the migration of prospective luteinizing hormone-releasing hormone (LHRH) from the olfactory placode to the hypothalamus (Norgren and Lehman 1991), of cortical interneurons from the ganglionic eminences (De Carlos et al. 1996; Anderson et al. 1997; Tamamaki et al. 1997) and of prospective olfactory interneurons from the subventricular or subependymal zone (SVZ) to the core of the olfactory bulb (Luskin 1993; Luskin and Boone 1994; Lois et al. 1996). Unlike the radial migration that depends on radial glia, this mode of migration relies on no glial scaffold and several possibilities have been hypothesized to account for the migration of neurons under such conditions. For example, it has been proposed that granular neuronal precursors migrate associated with axons to form the external granular layer of the cerebellum (Hynes et al. 1986), and that some of the cells tangentially migrating from the ganglionic eminences to the cortex use the incoming corticofugal fibers as a guide (Denaxa et al. 2001). It has also been shown that in the RMS neurons move rapidly along one another in unique chain formations independent of radial glia or axonal processes using their migrating neighbors to provide a foothold for their movement (Doetsch et al. 1997; Wichterle et al. 1997).

The tangential migration of prospective cerebellar and olfactory bulb granular neurons is followed by a final step of glial-assisted radial migration, away from and toward the pial surface, respectively. If it is assumed that the somatodendritic membrane derives from the basolateral domain of the primitive neuroepithelium, the (secondary) radial migration of cerebellar granular cells involves a reorientation if compared to the primary radial migration of projection neurons of the cerebral cortex and the radial migration of Purkinje cells (Lambert de Rouvroit and Goffinet 2001). However, that may not be the case with granular cells of the olfactory bulb which do not grow an axon (Shipley et al. 1995). It is interesting to note that interneurons of the cerebral cortex en route from the ganglionic eminence first "dive" towards the ventricular surface before reorienting and migrating radially towards the pial surface (Nadarajah et al. 2002).

The considerations above aim at raising the question of whether there are common molecular mechanisms underlying primary, secondary (or tertiary) radial migration, tangential migration along extrinsic fiber tracts or mediated by neuronal chains formed by the migrating neurons proper and other directional movements such as axonal pathfinding. In this respect, there is already evidence that some factors that control directional growth during tangential migration are similar to those that control the outgrowth of neuronal growth cones (for review, see Corbin et al. 2001).

In our studies we have identified and characterized a specific glycolipid localized to the membrane of neurons and glial cells that is expressed in regions of cell migration and neurite outgrowth in the developing and adult nervous system. This molecule is recognized by a monoclonal antibody named JONES and was identified as a ganglioside (Constantine-Paton et al. 1986; Blum and Barnstable 1987). In several studies we have shown that the distribution of this ganglioside is temporally and spatially correlated with neuronal migration and neurite outgrowth in the central and peripheral nervous system and we thus hypothesize that this ganglioside participates in critical stages of neuronal migration and neurite outgrowth. In this review, we will summarize our studies that provide evidence for the functional role of this glycolipid in neuronal motility.

2 Gangliosides and Cell Motility

In the vertebrate nervous system, the adhesive repertoire of cells participating in cell recognition and adhesive interactions during brain development comprises a number of well-defined cell-cell and cell-extracellular matrix receptor families (Takeichi 1991; Hynes and Lander 1992), and the identification of the selective components underlying neuronal cell deployment to specific laminae and nuclei is beginning to emerge. Most of these studies have focused on proteins, however, non-protein glycoconjugates such as glycolipids have also been suggested as important players in these processes.

Gangliosides are a large group of sialylated glycosphingolipids widely expressed in mammalian tissues and particularly abundant in the nervous system (Lloyd and Furukawa 1998). A dramatic feature of ganglioside biology is the differential expression during development and malignant transformation. From these observations no particular role can be assigned to them, but the general concept emerged that they may be important in development and tissue organization. A specific pattern of expression of gangliosides is determined by the balance of their biosynthesis and degeneration. Enzymological studies have shown that biosynthesis of gangliosides proceeds in a stepwise manner, starting from lactosyl ceramide and involves several glycosyltransferases (Fig. 1). Some of the genes coding for some of the glycosyltransferases have recently been cloned, but several questions are still under active investigation regarding how the levels of glycosyltransferases are regulated and determined (for review, see Lloyd and Furukawa 1998). An important modification of gangliosides is the O-acetylation of sialic acid residues on C-9, C-7 or C-9 and C-7 observed on restricted cells and tissues, but the enzymes responsible for these modifications (O-acetyl transferases) have been particularly difficult to clone. It has been suggested that O-acetylation of gangliosides may serve as a cellular recognition signal and also as tumor markers (Cheresh et al. 1984; Constantine-Paton et al. 1986). Sialidases are the key enzymes for ganglioside degradation and the plasma membrane-bound sialidase has been implicated not only in general catabolism of gangliosides, but also in modulation of cellular function, such as proliferation and differentia-

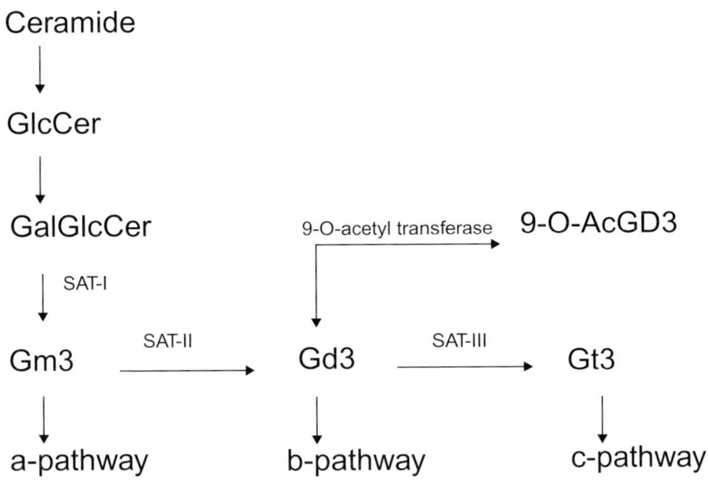

Fig. 1. Proposed pathway for the biosynthesis of gangliosides. *Glc* Glucose; *Cer* ceramide; *Gal* galactose; *SAT-I* GM3 synthase; *SAT-II* GD3 synthase; *SAT-III* GT3 synthase. The nomenclature is based on that of Svennerholm (1980)

tion (Reitzenstein et al. 2001). Recently, it has been shown that this enzyme is localized on membrane rafts, i.e., physically distinguishable microdomains associated with various functional membrane proteins (Miyagi et al. 1999; Kalka et al. 2001).

In the nervous system, the species and amounts of gangliosides undergo profound changes during development, suggesting that they may play fundamental roles in this process (Rosner 1994). In the vertebrate brain, the pattern of gangliosides changes from a relatively simple scheme, with GM3 and GD3 gangliosides predominating at earlier stages, to a more complex pattern with four gangliosides of the a- and b-series, GM1, GD1a, GD1b, and GT1b, that constitute the major species in the adult (for review, see Yu 1994). Furthermore, the accumulation of gangliosides within the neurons in ganglioside storage diseases results in extensive neurite outgrowth (Purpura 1978) and exogenously administered gangliosides accelerate the regeneration of neurons in the CNS in vivo and stimulates cellular differentiation with concomitant neurite sprouting and extension in vitro (Toffano et al. 1983; Skaper et al. 1985; Roisen et al. 1986). In addition, in preliminary studies, it was reported that GM1 appears to be beneficial in patients with spinal cord injury (Geisler et al. 2001). Another series of experiments aiming at elucidating the function of gangliosides have relied on blocking ganglioside biosynthesis with specific inhibitors or using cell lines that lack some or all gangliosides. For example, using specific blockers, it has been shown that glucosylceramide synthesis, the first glycosylation step of glycosphingolipid synthesis, is required for axonal growth in hippocampal neurons and for embryonic development (Schwarz et al. 1995; Yamashita et al. 1999). Recently, the genes involved in the biosynthesis of gangliosides have been cloned and genetically engineered mice or cells lacking most of the gangliosides have been produced (Takamiya et al. 1996; Kawai et al. 1998). Surprisingly, the mice and the cells appeared to be largely normal in their neuronal development. Inhibition of ganglioside synthesis in post-implantation mouse embryos, fish embryos and certain transformed neuron-like cells also showed that depletion of most of the gangliosides does not severely affect certain aspects of neuronal development (Fenderson et al. 1992; Takamiya et al. 1996; Li and Ladish 1997; Brigande et al. 1998; Kawai et al. 1998; Li et al. 1998). However, more recent studies revealed considerable functional and morphological abnormalities that cast a doubt as to the earlier conclusion that gangliosides are not important for neuronal development (Takamiya et al. 1996; Sheikh et al. 1999). For example, it was described that mice lacking complex gangliosides exhibit axonal degeneration and myelination defects (Sheikh et al. 1999; Chiavegatto et al. 2000).

The interpretation of many of the experiments described in the previous paragraphs is complicated mainly in those utilizing the addition in vitro or the administration in vivo of gangliosides to achieve a biological effect. This is due to the well known fact that cells have the tendency to convert the added gangliosides into both simpler or more complex species making it difficult to assign the observed effect to a specific ganglioside. In spite of that, from the observations summarized above, the general concept emerges that gangliosides may contribute to cell-cell recognition or cell-matrix interactions although no particular role can be assigned to individual gangliosides.

From experiments in several laboratories it has been shown that glycosphingolipids, in particular gangliosides, are emerging as components of membrane rafts, where they can mediate important physiological functions. Glycosphingolipids present in such microdomains have been used simply as markers of low-density membrane fraction and the functional role of glycosyl epitopes in such microdomains has received little attention. However, several studies have shown that glycosphingolipids present in these microdomains mediate cell adhesion and signal transduction events that affect cellular phenotypes and functions (for review, see Pande 2000). Such microdomains have been referred to as lipid rafts, detergent-resistant membranes, detergent-insoluble glycosphingolipid-enriched domains, or caveolae membranes (Harder and Simons 1997) and, recently, have been termed "glycosynapses" in analogy to "immunological synapses" – the membrane assembly of immunocyte adhesion and signaling (Hakomori 2002). Gangliosides are abundant in these microdomains and it has been shown that anti-ganglioside antibodies can immunoprecipitate GPI-anchored proteins (such as TAG-1), Src family kinases and caveolin (Iwabuchi et al. 1998; Katagiri et al. 1999; Kasahara et al. 2000). In addition, several studies have shown the association of gangliosides with integrins and their ability to modulate integrin function (Iwabuchi et al. 1998; Yates and Rampersaud 1998). In addition, several studies have also suggested that gangliosides can affect the distribution of GPI-anchored proteins in lipid rafts (Simons et al. 1999; Prinetti et al. 2000).

In summary, since gangliosides comprise a significant proportion of the total glycoconjugates that exist on the surface of neuronal cells, they have been assumed to be involved in many activities in which the cell surface plays a role. While there is no definitive evidence of their specific functions in the nervous system, a number of potential activities have been ascribed to them. Examples are: (1) modulation of growth factor receptors with intrinsic tyrosine kinases (for review, see Yates and Rampersaud 1998); (2) modulation of integrin function through their complex with tetraspanin CD9 (Ono et al. 2001); and (3) interactions with and activation of cytoplasmic signal transducers such as Src family kinases and small G proteins present in microdomains (Iwabuchi et al. 1998; Kasahara et al. 2000).

Our studies have focused on a specific ganglioside recognized by the monoclonal antibody JONES (Constantine-Paton et al. 1986). In our initial studies we have shown that the pattern of staining with this antibody in the developing rat nervous system correlates temporally and spatially with neuronal migration and neurite outgrowth (Mendez-Otero et al. 1988; Schlosshauer et al. 1988). The biochemical characterization of the antigen revealed that this mAb recognizes a ganglioside that migrates between GM1 and GM2 ganglioside standards. Further characterization using overlay assays on developed high-performance thin-layer chromatographic plates indicates that the epitope recognized by the JONES mAb is expressed on several regions of the developing CNS and PNS. In most regions examined, the epitope resides in two bands (Schlosshauer et al. 1988). Further analysis has revealed that the most abundant and frequent band is 9-O-acetyl GD3 (9-O-AcGD3; Fig. 2), a modification of GD3 ganglioside in which an acetyl ester is formed on the 9 position of the terminal sialic acid residue (Blum and Barnstable 1987). The particular epitope recognized by the JONES

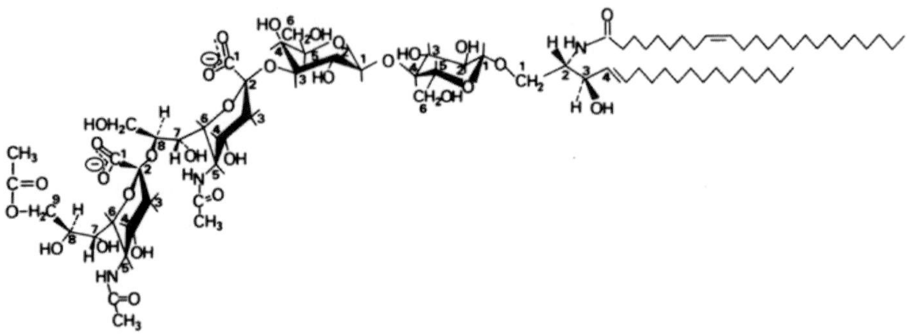

Fig. 2. Schematic diagram of 9-O-acetyl GD3. This ganglioside is formed by the addition of an acetyl group to the 9-position of the terminal sialic acid of GD3

mAb depends upon the acetyl group in the position 9 of the sialic acid and the immunoreactivity is abolished by mild base treatment since this treatment converts 9-O-AcGD3 to GD3 that is not recognized by the JONES mAb (Blum and Barnstable 1987; Bonafede et al. 1989; Ritter et al. 1990). This antigen is abundant in the nervous system, tumor cells derived from the neural crest and absent from most of the other tissues examined (Schlosshauer et al. 1888). However, the same epitope was described in the immune system and was designated CD60b (D. Mason et al. 2001). In the immune system, this antigen is expressed by a subpopulation of lymphocytes (Rieber and Rank 1994). The 9-O-acetyl GD3 is present in retinas of all mammalian species studied so far, but is not present in extracts of frog retinas (Mendez-Otero et al. 1992). At least two other mAbs have been described (mAb D1.1 and RB13–2) that recognize the same epitope in nervous tissue (Levine et al. 1984; Reinhardt-Maelicke et al. 1990).The slower migrating band has its immunoreactivity also abolished by mild base treatment and it has been suggested that this band might correspond to 9-O-acetyl GQ1c (Multani et al. 1988).

In the developing nervous system, we have shown that the expression pattern of this specific ganglioside correlates particularly with times of cell migration in the retina, superior colliculus, cerebellum, and telencephalon and in regions undergoing neurite extension, such as the developing optic tract, the white matter of the cerebellum, the dorsal roots, the trigeminal system, and olfactory nerve (Constantine-Paton et al. 1986; Mendez-Otero et al. 1988; Schlosshauer et al. 1988; Mendez-Otero et al. 1992; Mendez-Otero and Ramon-Cueto 1994; Mello and Mendez-Otero 1996; Mendez-Otero and Santiago 2001). This has led us to suggest that the function of 9-O-AcGD3 is to modulate motility either directly or by modifying the efficacy of some other component of the system. Association of 9-O-acetylated gangliosides with cell migration has been well characterized in cancer research where it was found that tumors arising from neural crest-derived tissues express high levels of 9-O-acetyl GD3. These gangliosides are concentrated in adhesion plaques and are involved in cell adhesion and migration in tumor cells (Cheresh et al. 1984, see also Lloyd and Furukawa 1998 for a review). In addition, it was shown that disialogangliosides co-immunoprecipitate with alfaVbeta3 integrin and GM1 co-immunoprecipitates with the epidermal growth factor receptor (EGFR; for review, see Hakomori 2002). Furthermore, it was also shown that cleavage of 9-O-acetyl groups by transgenic expression of influenza C virus hemagglutinin caused abnormalities in the development of the retinal layers in the mouse retina, suggesting a failure in cell migration concomitant with the absence of 9-O-acetyl GD3 (Varki et al. 1991). Recently, it has been shown that downregulation of 9-O-AcGD3 by stable transfection of O-acetylesterase cDNA and antisense vector against the GD3-synthase gene results in cell differentiation in melanoma cell lines (Birkle et al. 1999, 2000). Based on our observations and the reports in other systems, we made the hypothesis that 9-O-acetyl GD3 may play an important role in neuronal motility in the developing and adult nervous system. In the following sections, we will describe the evidence showing a functional role for this ganglioside in neuronal migration and neurite outgrowth.

3 Gangliosides and Glial-Guided Radial Migration

Neuronal radial migration occurs, in the majority of cases, on the processes of radial glial cells and has been studied extensively in the cortex and the cerebellum although similar arrangements between migrating neurons and specialized glia have been described in a number of other systems including the retina and the hippocampus (for reviews, see Rakic 1995; Hatten 1999). Recent studies have begun to provide a framework for the molecular mechanisms underlying this mode of migration. For instance, in the cerebral cortex, the radially migrating neurons associated with radial glia appear to rely for guidance on the *reelin* and *disabled* gene products, the reelin receptors ApoR2 and Vldr, and for movement on the products of the *lis1* and *doublecortin* genes. In addition, layers II through V require the activity of Cdk5 and its activator, p35, for proper movement. In the embryonic rat cerebellum, both the Purkinje cells and deep nuclear neurons migrate in a radial direction using a palisade of radial glial fibers as substrates and guides. As in the cerebral cortex, the *reelin* and *disabled* gene products as well as ApoR2 and Vldlr are important guidance components. Cdk5 activity is required for Purkinje cell, but not deep nuclear neuronal migration. Both cell types need Lis1, but doublecortin appears unnecessary (for reviews, see Gilmore and Herrup 1997; Gleeson and Walsh 2000; Walsh and Goffinet 2000; Lambert de Rouvroit and Goffinet 2001).

In the postnatal cerebellum, the postmitotic neurons in the external granular cell layer also migrate radially, but in this case away from the pia towards the internal granular cell layer. This migration is also associated with radial glia and this system has been extensively used as a model for studies of glial-guided migration. These studies have shown that neuronal cell migration is saltatory, alternating between active and inactive periods of cell movement and a more detailed analysis of the molecular mechanisms involved have suggested the involvement of several molecules in this migration (for reviews, see Pearlman et al. 1998; Hatten 1999). The pattern that has emerged from these studies is that multiple mechanisms such as recognition, adhesion, transmembrane signaling and cell motility are involved in glial-guided neuronal migration although the role of each molecular component that mediates each of these events is still unknown.

We have investigated the possible role of the 9-O-acetyl GD3 in the glial-guided neuronal migration using the developing rat cerebellum as a model. We have shown that the expression of this ganglioside is developmentally regulated in the developing cerebellum (Mendez-Otero et al. 1988). In E14–18 fetuses immunocytochemistry labeling with JONES mAb is present extending from the ventricular to the pial surface of the cerebellar anlage. Most of the immunoreactivity is distributed in a radially oriented pattern that corresponds to the radial migration of Purkinje cell precursors migrating from the ventricular zone towards the pial surface. Immunoreactivity, however, is present in the rhombic lip and in the developing external granular cell layer corresponding to the subpial tangential migration of these cells. During the postnatal period in the rat, the cells in the external granular cell layer proliferate and migrate radially away from the pia to form the internal granular cell layer. Immunocytochemistry reveals a radially oriented pattern of JONES binding extending from the external granular cell layer towards the presumptive granular cell layer. Electron microscopic immunocytochemistry revealed that around the peak of cerebellar neuronal migration, 9-O-acetyl GD3 was localized at the contact sites between migrating granule cells and radial glia in the external granular cell layer and prospective molecular layer (Santiago et al. 2001). We have also observed

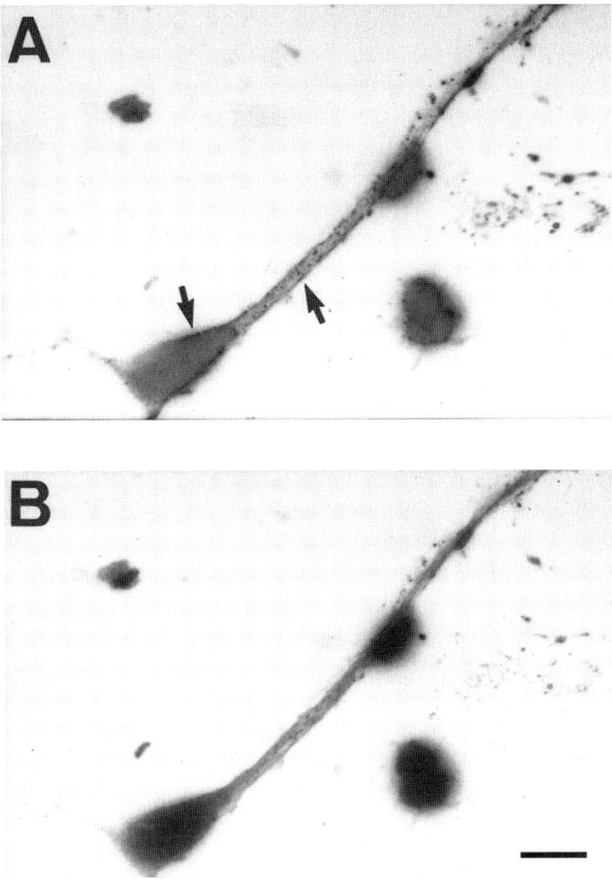

Fig. 3A,B. Correlation of 9-O-acetyl GD3 expression with radial cell migration. Cerebellar microculture double stained with JONES mAb (**A**) and glial fibrillary acidic protein-GFAP (**B**). Note the punctate staining outlining the glial cell body and processes (*arrows*). Calibration bar: 10 μm

that both cultured neurons from rat cerebella when in contact with glial cells that support cell migration (radial glial cells) and glial cells themselves express 9-O-AcGD3 (Fig. 3; see also Mendez-Otero and Constantine-Paton 1990). On the other hand, glial cells with a stellate morphology, even when in contact with granule neurons do not express this antigen (Mendez-Otero and Constantine-Paton 1990). To test the functional role of 9-O-acetyl GD3 in neuronal migration we have used cerebellar explants and cerebellar slices. We have shown that the JONES mAb blocks the migration of neurons in a dose-dependent manner suggesting that 9-O-AcGD3 is involved in the radial migration of presumptive granule cells in the developing cerebellum (Santiago et al. 2001). These results further support our view that this molecule is involved in cell migration. We specifically tested the role of this ganglioside in gliophilic (Bergmann-like glia) radial migration of granule cells, but it is important to mention that this molecule is also associated with Purkinje cell migration that takes place at embryonic ages and with the tangential migration of precursors from the rhombic lip to form the external granule cell layer (Mendez-Otero et al. 1988). The immunoreactivity for this ganglioside is also present in the embryonic telencephalon showing a radial organization correlated with radial migration in this region (Mendez-Otero et al. 1988). Blockage of the ganglioside with JONES antibody arrests neuronal migration on tissue slices of embryonic telencephalon (Hedin-Pereira et al. 1997).

Our traditional view of radial glial cells is that of specialized non-neuronal cells of the astroglial lineage found in the developing CNS that serve as a scaffold to support and direct neurons during their migration. This view is now changing due to the emerging evidence showing that these cells have a more dynamic and diverse role. Recent series of studies have provided strong support for their role as precursor cells in the ventricular zone that generate cortical neurons and glia, in addition to providing migration guidance (Noctor et al. 2001; Parnavelas and Nadarajah 2001). It would be important to test whether JONES-labeled radial glia (Fig. 3), in addition to their role in neuronal migration, can also generate neurons and glia.

4 Gangliosides and Tangential Migration

It is usually believed that glial-directed neuronal migration, whether primary as in the cerebral cortex and Purkinje cells, or secondary as in the case of the cerebellar granule cells has been relatively well studied, whereas much less is known about tangential migration (for reviews, see Corbin et al. 2001; Maricich et al. 2001). Tangentially migrating neurons are found in the cortex, the cerebellum and in the rhombencephalon and they migrate perpendicularly to the glial scaffolding (Ono and Kawamura 1989; Ryder and Cepko 1994; Tamamaki et al. 1997; Corbin et al. 2001; Komuro et al. 2001). Recently, some of the cues that guide tangential migration have started being revealed and the picture that is emerging is that the same families of signals involved in the directional guidance of developing growth cones may also guide the translocation of cell bodies (Yee et al. 1999; Alcantara et al. 2000; Denaxa et al. 2001). In general, it has been proposed that tangentially migrating cells may move by using many of the same molecules used by growing axons such as N-cadherins, NCAM, TAG-1 and NrCAM (Kyriakopolou et al. 2002). Alternatively, these migrating cells may use as yet uncharacterized modes of migration, and/or molecules, to promote their movement.

The migration of interneurons from the anterior subventricular zone (SVZ) to the olfactory bulb provides a useful system to study the cellular and molecular mechanisms that regulate tangential neural migration. These neuronal precursors migrate through a distinct pathway within the SVZ called the rostral migratory stream (RMS), and this migration occurs throughout life. Migrating cells in the RMS are closely associated with one another, forming elongated cell aggregates resembling chains. Several extracellular molecules are involved in neuronal migration in the RMS. Molecular studies have shown that, in the particular case of the rostral migratory stream, the mechanism of migration involves the sialylated form of NCAM (neural cell adhesion molecule) and is guided, in part, by negative chemotropism (Hu et al. 1996). Mice lacking the polysialylated form of the neural cell adhesion molecule (PSA-NCAM) have reduced olfactory bulbs, presumably attributable to reduced migration and enzymatic removal of polysialic acid (PSA) from NCAM in SVZ explants resulting in decreased migration (Chazal et al. 2000). In addition, altering Eph receptor tyrosine kinase signaling appears to disrupt the proliferation and migration of neuroblasts in the SVZ. Finally, Slit, a secreted protein that interacts with Robo receptors has been proposed to act as a repellent for migrating SVZ cells, because cells from SVZ explants migrate away from localized sources of Slit (for review, see Corbin et al. 2001). Recently, it was described that astrocytes in the RMS release a soluble protein called MIA, for migration-inducing activity, which induces the migration of olfactory bulb interneuron precursors, increasing the number of migrating cells and the distance they move (H.A. Mason et al. 2001). In view of the abundant evidence for a role of 9-O-acetyl GD3 in directional movements of neuronal soma or processes, we have made an immunohistochemical analysis of their expression along the route of tangential migrations of some olfactory bulb prospective interneurons in the developing and adult rats. We have found that this ganglioside is highly expressed in the lateral ventricle SVZ and along the route of tangential migration (RMS) into the olfactory bulb during development (Mendez-Otero and Cavalcante 1996).

In new-born rats we have shown a very strong expression of 9AcGD3 around the lateral ventricle. In parasaggital sections, these stained regions descend towards the olfactory bulb forming a continuous,

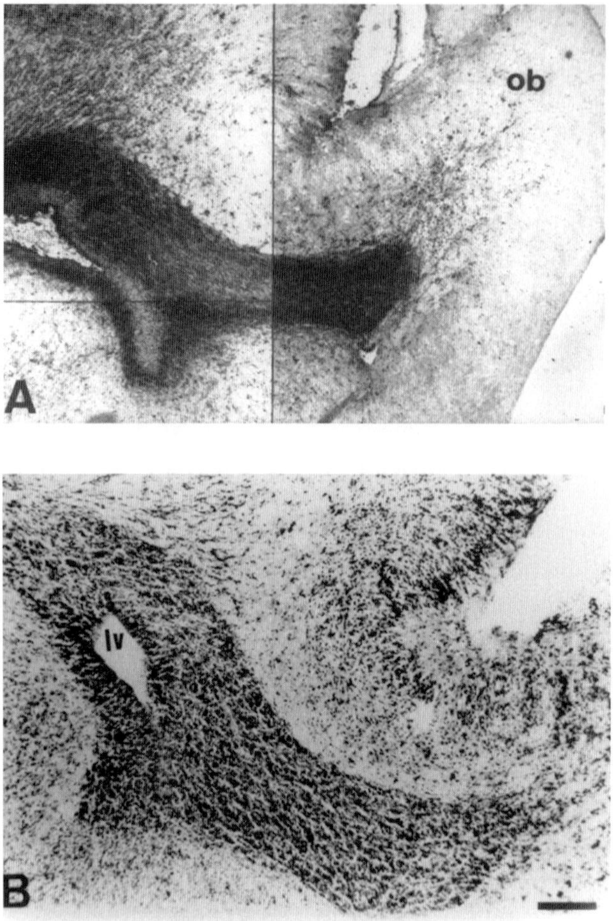

Fig. 4A,B. Expression of 9-O-AcGD3 is correlated with tangential cell migration. **A** Photomicrograph of a parasagittal section through the telencephalon of a newborn rat stained with the JONES mAb. Note the intense staining around the lateral ventricle (*lv*) extending towards the olfactory bulb (*ob*), thus, including the subventricular zone plus the rostral migratory stream. Due to a slight obliquity of the section, the intrabulbar subventricular zone is not seen in this plane of section and labeling assumes a fan-shaped distribution due to secondary radial migration of interneurons. **B** Photomicrograph of an adjacent section stained with toluidin blue. Calibration bar: 100 μm

compact strip of immunoreactivity corresponding to the RMS (Fig. 4). A similar, although less intense staining is found in the postnatal day 7 animal and there is a progressive reduction in the intensity of the reaction in the subsequent days. In the adult, we have found staining around the lateral ventricle and in chains of cells in the RMS region. In a few sections, individual stained cells are observed and they display a morphology similar to that of migrating cells. Notwithstanding the scarcity of staining, this is one of the few regions where expression of 9-O-acetyl GD3 is found in the adult nervous system. Interestingly, this region has been described as one of the regions in which neurogenesis remains in the adult brain and from which stem cells can be isolated in the adult brain. It would be interesting to investigate whether the stem cells isolated from adult brains also express 9-O-acetyl GD3.

We have investigated the functional role of 9-O-acetyl gangliosides in tangential migration using explants of the SVZ region as a model. We have found that migrating chains similar to the ones formed in vivo are also seen in this in vitro system. The migrating chains from the SVZ explants express 9-O-acetyl GD3 which is distributed in a punctate manner in individual cells. 9-O-acetyl GD3 is also present in migrating chains that form in the absence of radial glia, typical of the neuronophilic chain migration of the SVZ. Furthermore, immunoblockage of 9-O-GD3 arrested neuronal migration. These data indicate that 9-O-acetyl gangliosides may participate in neuronophilic as well as gliophilic migration (Miyakoshi et al. 2001a, b).

5 Gangliosides and Neurite Outgrowth

While it has been clear for a number of years that the neuronal growth cone is involved in sensing and transducing environmental signals, the molecular mechanisms of growth cone guidance have only recently begun to be understood (for reviews, see Walsh and Doherty 1997; Gallo and Letourneau 1999). During development, neuronal growth cones interact with physical and chemical cues in their environment. These interactions guide the growth cones along specific pathways to their appropriate targets. Many factors such as soluble or substratum-bound growth factors and components of the extracellular matrix provide favorable environments that allow or promote motility of growth cones and neurite elongation (Dodd and Jessell 1988; Bixby and Harris 1991; Garcia-Abreu et al. 2000). Repulsive and inhibitory factors have also been described and in concert with molecules favorable to axon growth, may participate in guidance of growth cones (for reviews, see Gallo and Letourneau 1999; Mueller 1999). Gangliosides, in particular disialogangliosides, have been implicated in adhesion systems and in cell to substratum adhesion of both melanoma and neuroblastoma cells (Cheresh et al. 1984). Indeed, a number of separate lines of evidence support the hypothesis that these molecules modulate the activity of the integrin family of cell surface adhesion molecules (for review, see Pande 2000) and induce selectively the synthesis of high molecular weight microtubule-associated protein 2 (MAP2 – Ferreira et al. 1990). Furthermore, numerous studies employing a variety of models have reported that ganglioside treatment will increase rates of neuritogenesis both in vitro and in vivo, and axonal regeneration and sprouting following brain trauma to the central or peripheral nervous system. In addition, antibodies to gangliosides have been shown to inhibit goldfish optic nerve regeneration (Finnegan et al. 1992). These observations suggested a role for gangliosides in neurite outgrowth.

In our studies of the expression of 9-O-acetyl GD3 in the developing nervous system we have found that, in addition to the pattern of staining coincident with neuronal migration, these gangliosides were also observed along pathways of axon extensions such as the optic tract, the olfactory nerve and the central and peripheral processes of dorsal root ganglia. This pathway staining is only observed while large numbers of growth cones are present in these positions and that disappears from most of the pathways studied as soon as the axons reach their target regions (Mendez-Otero et al. 1988; Mendez-Otero and Ramon-Cueto 1994). One interesting exception to this generalization, however, is the olfactory system (Mendez-Otero and Ramon-Cueto 1994). In this system we have shown that the expression of 9-O-acetylGD3 begins as early as embryonic day 13 where the olfactory epithelium and the migratory mass were intensely stained. At embryonic day 19, the immunoreactivity has disappeared from the olfactory epithelium, but remained in a few fascicles and some glomeruli of the olfactory bulb in the new-born and adult. We conclude that the expression of 9-O-acetyl GD3 by olfactory axons and/or migrating cells may facilitate axonal outgrowth during development and might be involved in the formation of new glomeruli in the mature olfactory bulb (Mendez-Otero and Ramon-Cueto 1994).

Interesting results were also reported by Bahr and Schlosshauer (1989) while studying the regeneration of the optic nerve in the adult rat. In the adult visual system, 9-O-acetyl gangliosides are found only in the optic nerve fiber layer and in the optic nerve head and this staining stops abruptly as the nerve leaves the orbit (Mendez-Otero et al. 1988). In animals that have undergone an optic nerve axotomy or received an autologous sciatic nerve graft, a transient increase in JONES ganglioside expression in the retina was observed. The highest levels of ganglioside expression were observed 5 days after surgery – contemporaneous with the maximum regenerative response of the adult rat retina after crush lesions. This study has also shown that the ganglioside is expressed by retinal glial cells (Bahr and Schlosshauer 1989).

Recently, we have investigated whether 9-O-acetyl gangliosides are also re-expressed during the regeneration of sciatic nerves that were submitted to crush. We have found an upregulation of 9-O-AcGD3 expression starting 5 days after crush, reaching a maximum at day 7, and decreasing afterwards to reach the level of the control animals at day 10 (R. Mendez-Otero et al., unpubl. data). These results suggested to us that 9-O-GD3 may play a functional role in neurite outgrowth during development and regeneration. Searching for an adequate model to study the possible role of 9-O-acetyl GD3 in neurite outgrowth, we found that embryonic dorsal root ganglia (DRG) neurons and their processes either in vivo (Mendez-Otero et al. 1988), in explant cultures and as dissociated cells, present heavy immunoreactivity to the antibody against this ganglioside. We then plated DRG explants from embryonic day 16 on laminin, collagen or fibronectin. Neurites grow out of the explants and express high amounts of 9-O-acetyl GD3 in all the substrates (Fig. 5). We then investigated the functional role of this ganglioside on neurite extension using embryonic DRG explants grown on laminin substrates as a model. In these experiments, the behavior of individual growth cones was recorded using a time-lapse video-enhanced imaging system before and after the addition of antibodies that recognize specific gangliosides known to be expressed on these growth cones. Using this system it was possible to demonstrate that the advance of growth cones on laminin was halted in the presence of JONES mAb. The onset of the effect was rapid and signaled by an immediate cessation of elongation, a loss of lamellipodia and a retrieval of axoplasm (Mendez-Otero and Friedman 1996). Blockage of 9-O-acetyl GD3 also induces the appearance of lateral spikes along the neurites in. addition to the effect on the growth cone. We have shown that microfilaments and microtubular rearrangements accompany these structural modifications (Araujo et al. 1997). Our results suggest that 9-O-acetyl GD3 may play an important role on the extension of growth cones and consequently influence navigation and pathway finding during development and regeneration.

An important question is whether gangliosides have a direct role on axonal elongation or whether this effect is mediated indirectly through interactions with cell surface receptors for extracellular molecules. Since gangliosides have been shown to colocalize with focal adhesion sites and regulate the function of integrin receptors for extracellular matrix proteins in several cell types, we investigated whether this is also the case in neuronal growth cones. Using immunocytochemistry and con-

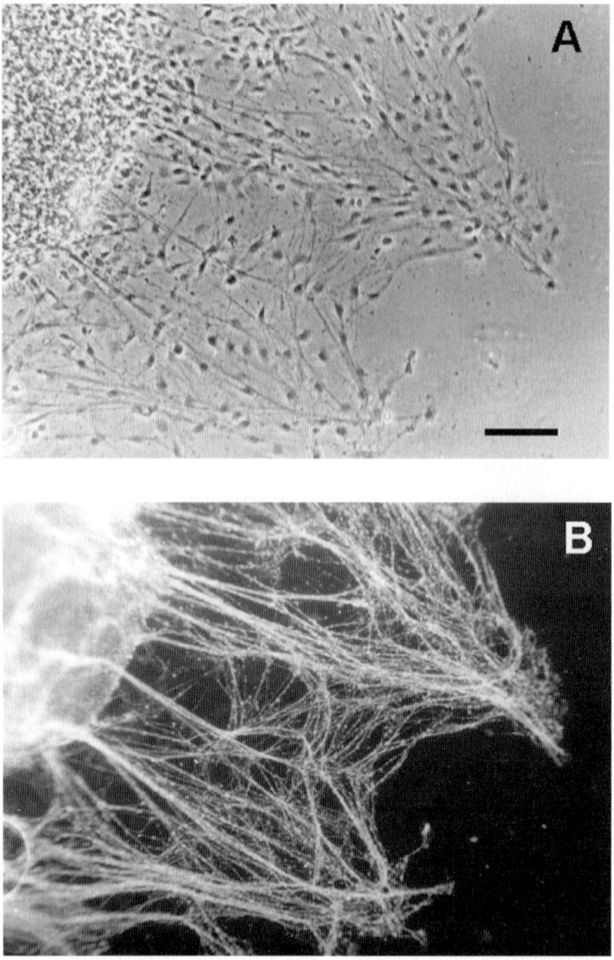

Fig. 5A,B. Expression of 9-O-acetyl ganglioside on neurites from dorsal root ganglion explants. **A** Phase contrast and **B** fluorescence photomicrographs of an explant of embryonic day 16 dorsal root ganglion plated on collagen substrate and stained with Jones mAb. Neurites, but not glia, show a high expression of 9-O-acetyl GD3. Calibration bar: 100 μm

focal microscopy, we examined whether this ganglioside is localized in point contacts in neuronal growth cones. Point contacts were identified by immunoreaction with proteins such as vinculin, known to be associated with these structures in growth cones (Renaudin et al. 1999). Our observations indicate that 9-O-acetyl GD3 is specifically associated with vinculin in point contacts of growth cones suggesting a possible role of this specific ganglioside in the modulation of these contacts during neurite outgrowth (Mendez-Otero et al., unpubl. data).

6 Conclusion and Perspectives

The mechanisms by which ganglioside expression leads to neuronal migration and neurite outgrowth are still far beyond been completely identified. However, based on the data reviewed in the previous sections and also in several findings from different laboratories it is possible to suggest at least three possible mechanisms to explain the role of 9-O-acetyl GD3 in these phenomena.

Gangliosides have been characterized by their ability to interact with key transmembrane receptors or signal transducers involved in cell adhesion and signaling. The mechanisms of ganglioside-receptor interactions, however, are still unknown. According to the mechanisms illustrated in Fig. 6I, it is likely that specific gangliosides could interact with broadly distributed protein-based adhesion systems and favored subsets of moving cells and/or processes that express these molecules. Examples of such mechanism are the described interactions of gangliosides with integrin receptors (Cheresh and Klier 1986; Cheresh et al. 1986; Probstmeier and Pesheva 1999) and with TAG-1 (Kasahara et al. 2000). Several possibilities have been suggested since gangliosides comprise a lipid moiety, ceramide, which is embedded in the outer leaflet of the bilayer, and a sugar part that extends mainly in the outer leaflet of cell membranes where they interact with a number of receptors. For example, the sialic acid residues of gangliosides are known to be efficient chelators of divalent cations. One possibility, therefore, is that gangliosides may serve to modulate the availability of divalent cations to a number of different calcium- or magnesium-dependent adhesion systems and consequently regulate ligand binding (Cheresh et al. 1986). In addition, gangliosides may regulate lateral diffusion and assembly of signaling complexes in membrane microdomains and interfere with the function of these proteins as it has been proposed for integrin in focal adhesion points and TAG-1 (Horwitz and Parsons 1999; Katz et al. 2000). Another possible model is based on the homophilic interaction between gangliosides located at different cell membranes (Fig. 6-II). A similar model has been proposed to explain the involvement of NCAM in regulating neural cell adhesion and axon growth. Finally, it has also been shown in several systems that selectins and/or galectins can function as receptors for specific gangliosides (Kopitz et al. 1998) and this interaction could trigger a cascade of reactions in both cells (Fig. 6III). It is also important to mention that specific receptors for gangliosides, including myelin-associated glycoprotein, have been identified in the brain (Yang et al. 1996; Vyas and Schnaar 2001). In conclusion, the

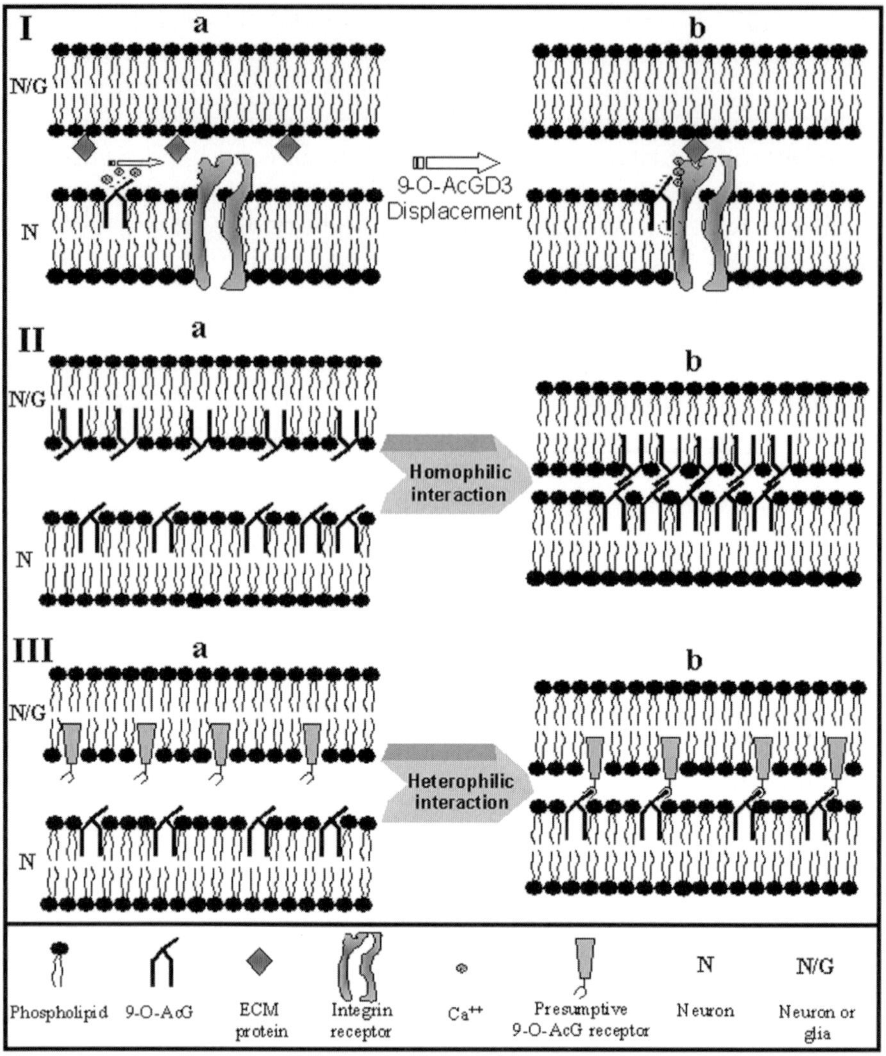

Fig. 6. Schematic diagrams suggesting three possible mechanisms for 9-O-acetyl GD3 (9-O-AcG) action in neuronal motility. **I** The ganglioside modulates the integrin through a Ca^{2+}-dependent mechanism and/or by a direct interaction between the ceramide moiety of the ganglioside and the integrin transmembrane helix. In *a*, the integrin is inactive and there is no recognition of extracellular matrix (ECM) proteins. In *b*, gangliosides in the adjacent membrane become laterally packed around the integrin, providing an optimal Ca^{2+}-enriched microenvironment for the activation and recognition of ECM proteins. In addition, the gangliosides can interact directly with the integrin through a conserved lysine located inside a 23 amino acids sequence at the carboxy-terminal of the integrin transmembrane helix. **II** Homophilic interaction between gangliosides on different cell membranes. In *a*, two cells containing 9-O-AcG randomly spread over the membranes. In *b*, the cells are interacting in a homophilic manner through a carbohydrate-carbohydrate interaction of the gangliosides. **III** Heterophilic interaction between gangliosides and a potential specific receptor that could trigger a specific cascade of reactions in both cells. (Reproduced with permission from Mendez-Otero and Santiago 2001)

ganglioside 9-O-acetyl GD3 could provide a new cell signaling mechanism in glial-guided neuronal migration in the developing nervous system. Moreover, it is important to note that most neurons migrate by extension of a leading, neurite-like process and that 9-O-acetyl GD3 has been implicated in both neuronal migration and neurite outgrowth. The potential roles for this ganglioside in identical mechanisms for neuronal migration and neurite outgrowth suggest an emerging framework in which glycolipids are involved in cell movement in general. Nevertheless, many important questions remain to be investigated concerning how this specific ganglioside exerts its role in neuronal motility.

Acknowledgements. Our laboratories are supported by PRONEX, CNPq, and FAPERJ grants and by a grant from the Ministry of Science and Technology (MCT) to the Millennium Institute for Tissue Bioengineering, Brazil. The technical assistance of Felipe Marins and Sergio L. Carvalho is gratefully acknowledged.

References

Aicardi J (1994) The place of neuronal migration abnormalities in child neurology. Can J Neurol Sci 21:185–193

Alcantara S, Ruiz M, De Castro F, Soriano E, Sotelo C (2000) Netrin 1 acts as an attractive or as a repulsive cue for distinct migrating neurons during the development of the cerebellar system. Development 127:1359–1372

Altman J (1969a) Autoradiographic and histological studies of postnatal neurogenesis. IV. Cell proliferation and migration in the anterior forebrain, with special reference to persisting neurogenesis in the olfactory bulb. J Comp Neurol 137:433–458

Altman J (1969b) Autoradiographic and histological studies of postnatal neurogenesis. III. Dating the time of production and onset of differentiation of cerebellar microneurons in rats. J Comp Neurol 136:269–294

Anderson AS, Eisenstat DD, Shi L, Rubenstein JLR (1997) Interneuron migration from basal forebrain to neocortex: dependence on Dlx genes. Science 278:474–476

Araujo H, Menezes MA, Mendez-Otero R (1997) Blockage of 9-O-acetyl gangliosides induces depolymerization in growth cones and neurites. Eur J Cell Biol 72:202–213

Bahr M, Schlosshauer B (1989) JONES ganglioside expression on retinal glia increases after axotomy. J Neurocytol 18:553–563

Birkle S, Ren S, Slominski A, Zeng G, Gao L, Yu RK (1999) Down-regulation of the expression of O-acetyl-GD3 by the O-acetylesterase cDNA in hamster melanoma cells: effects on cellular proliferation, differentiation, and melanogenesis. J Neurochem 72:954–961

Birkle S, Gao L, Zeng G, Yu RK (2000) Down regulation of GD3 ganglioside and its O-acetylated derivative by stable transfection with antisense vector against GD3-synthase gene expression in hamster melanoma cells: effects on cellular growth, melanogenesis, and dendricity. J Neurochem 74:547–554

Bixby JL, Harris WA (1991) Molecular mechanisms of axon growth and guidance. Annu Rev Cell Biol 7:117–159

Blackburn CC, Swank-Hill P, Schnaar RL (1986) Gangliosides support neural retina cell adhesion. J Biol Chem 261:2873–2881

Blum AS, Barnstable CJ (1987) O-acetylation of a cell-surface carbohydrate creates discrete molecular patterns during neuronal development. Proc Natl Acad Sci USA 84:8716–8720

Bonafede DM, Macala LJ, Constantine-Paton M, Yu RK (1989) Isolation and characterization of ganglioside 9-O-acetyl-GD3 from bovine buttermilk. Lipids 24:680–684

Bourrat F, Sotelo C (1988) Migratory pathways and neuritic differentiation of the inferior olivary neurons in the rat embryo. Axonal tracing study using the in vitro slab technique. Brain Res 467:19–37

Brigande JV, Platt FM, Seyfried TN (1998) Inhibition of glycosphingolipid biosynthesis does not impair growth or morphogenesis of the postimplantation mouse embryo. J Neurochem 70:871–882

Cameron RS, Rakic P (1994) Identification of membrane proteins that comprise the plasmalemmal junction between migrating neurons and radial glial cells. J Neurosci 14:3139–3155

Chazal G, Durbec P, Jankovski A, Rougon G, Cremer H (2000) Consequences of neural cell adhesion molecule deficiency on cell migration in the rostral migratory stream of the mouse. J Neurosci 20:1446–1457

Cheresh DA, Klier FG (1986) Disialoganglioside GD2 distributes preferentially into substrate associated microprocesses on human melanoma cells during their attachment to fibronectin. J Cell Biol 102:1887–1897

Cheresh DA, Varki AP, Varki NM, Stallcup WB, Levine J, Reisfeld RA (1984) A monoclonal antibody recognizes an O-acetylated sialic acid in a human melanoma-associated ganglioside. J Biol Chem 259:7453–7459

Cheresh DA, Pierschbacher MD, Herzig MA, Mujoo K (1986) Disialogangliosides GD2 and GD3 are involved in the attachment of human melanoma and neuroblastoma cells to extracellular matrix proteins. J Cell Biol 102:688–696

Chiavegatto S, Sun J, Nelson RJ, Schnaar RL (2000) A functional role for complex gangliosides: motor deficits in GM2/GD2 synthase knockout mice. Exp Neurol 166:227–234

Constantine-Paton M, Blum AS, Mendez-Otero R, Barnstable C (1986) A cell surface molecule distributed in a dorso-ventral gradient in the perinatal rat retina. Nature 324:459–462

Corbin JG, Nery, S, Fishell G (2001) Telencephalic cells take a tangent: non-radial migration in the mammalian forebrain. Nature Neurosci 4:1177–1182

de Carlos JA, Lopez-Mascarade L, Valverde F (1996) Dynamics of cell migration from the lateral ganglionic eminence in the rat. J Neurosci 16:6146–6156

de Diego I, Kyriakopoulou K, Karagogeos D, Wassef M (2002) Multiple influences on the migration of precerebellar neurons in the caudal medulla. Development 129:297–306

Denaxa M, Chan C-H, Schachner M, Parnavelas JG, Karagogeos, D (2001) The adhesion molecule TAG-1 mediates the migration of cortical interneurons from the ganglionic eminence along the corticofugal fiber system. Development 128:4635–4644

Dodd J, Jessell TM (1988) Axon guidance and the patterning of neuronal projections in vertebrates. Science 242:692–699

Doetsch F, Garcia-Verdugo JM, Alvarez-Buylla A (1997) Cellular composition and three dimensional organization of the subventricular germinal zone in the adult mammalian brain. J Neurosci 17:5046–5061

Edmondson J, Liem R, Kuster J, Hatten ME (1988) Astrotactin: a novel neuronal cell surface antigen that mediates neuron-astroglial interactions in cerebellar microcultures. J Cell Biol 106:505–517

Fenderson BA, Ostrander GK, Hausken Z, Radin NS, Hakomori S (1992) A ceramide analogue (PDMP) inhibits glycolipid synthesis in fish embryos. Exp Cell Res 198:362–366

Ferreira A, Busciglio J, Landa C, Caceres A (1990) Ganglioside-enhanced neurite growth: evidence for a selective induction of high molecular weight MAP-2. J Neurosci 10:293–302

Finnegan SG, Lemmon VP, Koenig E (1992) Characterization of the retraction response of goldfish retinal ganglion cell axons induced by monoclonal antibody 8A2 in vitro. Cell Mot Cytoskeleton 23:279–301

Fishell G, Hatten ME (1991) Astrotactin provides a neuronal receptor system for CNS migration. Development 113:755–765

Gage FH, Kempermann G, Palmer TD, Peterson DA, Ray J (1998) Multipotent progenitor cells in the adult dentate gyrus. J Neurobiol 36:249–266

Gallo G, Letourneau PC (1999) Axon guidance: a balance of signals sets axons on the right track. Curr Biol 9:R490–R492

Garcia-Abreu J, Mendes FA, Onofre GR, Freitas MS, Silva LCF, Moura Neto V, Cavalcante LA (2000) Contribution of heparan sulfate to the non-permissive role of the midline glia to the growth of midbrain neurites. Glia 29:260–272

Geisler FH, Coleman WP, Grieco G, Poonian D (2001) The syngen (r) multicenter acute spinal cord injury study. Spine 15:87–98

Gilmore E, Herrup K (1997) Cortical development: layers of complexity. Curr Biol 7:R231–R234

Gleeson J, Walsh C (2000) Neuronal migration disorders: from genetic diseases to developmental mechanisms. Trends Neurosci 23:352–359

Hakomori SI (2002) The glycosynapse. Proc Natl Acad Sci USA 8:225–232

Harder T, Simons K (1997) Caveolae, DIGs, and the dynamics of sphingolipid-cholesterol microdomains. Curr Opin Cell Biol 9:534–542

Hatten ME (1999) Central nervous system neuronal migration. Annu Rev Neurosci 22:511–539

Hedin-Pereira C, deMoraes ECP, Lent R, Mendez-Otero R (1997) Immunoblocking 9-O-acetylated gangliosides arrests cell migration in the developing cerebral cortex. Soc Neurosci Abstr 23:871

Horwitz AR, Parsons JT (1999) Cell migration – movin' on. Science 286:1102–1103

Hu H, Tomasiewics H, Magnuson T, Rutishauser U (1996) The role of polysialic acid in migration of olfactory bulb interneuron precursors in the subventricular zone. Neuron 16:735–743

Hynes RO, Lander AD (1992) Contact and adhesive specificities in the associations, migrations, and targeting of cells and axons. Cell 68:303–322

Hynes RO, Patel R, Miller RH (1986) Migration of neuroblasts along preexisting axonal tracts during prenatal cerebellar development. J Neurosci 6:867–876

Iwabuchi K, Handa K, Hakomori S (1998) Separation of "glycosphingolipid signaling domain" from caveolin-containing membrane fraction in mouse melanoma B16 cells and its role in cell adhesion coupled with signaling. J Biol Chem 273:33766–33773

Kalka D, von Reitzenstein C, Kopitz J, Cantz M (2001) The plasma membrane ganglioside sialidase cofractionates with markers of lipid rafts. Biochem Biophys Res Commun 18:989–993

Kasahara K, Watanabe K, Takeuchi K, Kaneko H, Oohira A, Yamamoto T, Sanai Y (2000) Involvement of gangliosides in glycosyl-phosphatidylinositol-anchored neuronal cell adhesion molecule TAG-1 signaling in lipid rafts. J Biol Chem 275:34701–34709

Katagiri YU, Mori T, Hakajima H, Katagiri C, Taguchi T, Takeda T, Kiyokawa N, Fujimoto J (1999) Activation of Src family kinase yes induced by Shiga toxin binding to globotriaosyl ceramide (Gb3/CD77) in low density, detergent-insoluble microdomains. J Biol Chem 273:35278–3582

Katz B-Z, Zamir E, Bershadsky A, Yamada KM, Geiger B (2000) Physical state of the extracellular matrix regulates the structure and composition of cell-matrix adhesions. Mol Biol Cell 11:1047–1060

Kawai H, Sango K, Mullin KA, Proia RL (1998) Embryonic stem cells with a disrupted GD3 synthase gene undergo neuronal differentiation in the absence of b-series gangliosides. J Biol Chem 273:19634–19638

Komuro H, Yacubova E, Yacubova E, Rakic P (2001) Mode and tempo of tangential migration in the cerebellar external granular layer. J Neurosci 15:527–540

Kopitz J, Reitzenstein C, Burchert M, Cantz M, Gabius HJ (1998) galectin-1 is a major receptor for ganglioside GM1, a product of the growth-controlling activity of a cell surface ganglioside sialidase, on human neuroblastoma cells in culture. J Biol Chem 273:11205–11211

Kyriakopolou K, de Diego I, Wassef M, Karagogeos D (2002) A combination of chain and neurophilic migration involving the adhesion molecule TAG-1 in the caudal medulla. Development 129:287–296

Lambert de Rouvroit C, Goffinet AM (2001) Neuronal migration. Mech Dev 105:47–56

Levine JM, Beasly L, Stallcup WB (1984) The D1.1 antigen : a cell surface marker for germinal cells of the central nervous system. J Neurosci 4:820–831

Li E, Ladish S (1997) Inhibition of endogenous ganglioside synthesis does not block neurite formation by retinoic acid-treated neuroblastoma cells. J Biol Chem 272:1349–1354

Li R, Kong Y, Ladish S (1998) Nerve growth factor-induced neurite formation in PC12 cells dependent of endogenous cellular gangliosides. Glycobiology 8:597–603

Lloyd KO, Furukawa K (1998) Biosynthesis and functions of gangliosides: Recent advances. Glycoconj J 7:627–636

Lois C, Garcia-Verdugo J-M, Alvarez-Buylla A (1996) Chain migration of neuronal precursors. Science 271:978–981

Luskin MB (1993) Restricted proliferation and migration of postnatally generated neurons derived from the forebrain subventricular zone. Neuron 11:173–189

Luskin MB, Boone MS (1994) Rate and pattern of migration of lineally related olfactory bulb interneurons generated postnatally in the subventricular zone of the rat. Chem Senses 19:695–714

Maricich SM, Gilmore EC, Herrup K (2001) The role of tangential migration in the establishment of mammalian cortex. Neuron 31:175–178

Mason D, Andre P, Bensussan A, Buckley C, Civin C, Clark E, Haas M, Goyert S, Hadam M, Hart D, Horejsi V, Meuer S, Morrissey J, Schwartz-Albiez R, Shaw S, Simmons D, Uguccioni M, van der Schoot E, Vivier E, Zola H (2001) CD antigens 2001: aims and results of HLDA workshops. Stem Cells 19:556–562

Mason HA, Ito S, Corfas G (2001) Extracellular signals that regulate the tangential migration of olfactory bulb neuronal precursors: inducers, inhibitors, and repellents. J Neurosci 21:7654–7663

Mello LEAM, Mendez-Otero R (1996) Expression of 9-O-acetylated gangliosides in the rat hippocampus. Neurosci Lett 213:17–20
Mendez-Otero R, Cavalcante LA (1996) Expression of 9-O-acetylated gangliosides is correlated with tangential cell migration. Neurosci Lett 204:97–100
Mendez-Otero R, Constantine-Paton M (1990) Granule cell induction of 9-O-acetyl gangliosides on cerebellar glia in microcultures. Dev Biol 138:400–409
Mendez-Otero R, Friedman JE (1996) Role of acetylated gangliosides on neurite extension. Eur J Cell Biol 71:192–198
Mendez-Otero R, Ramon-Cueto A (1994) Expression of 9-O-acetylated gangliosides during development of the rat olfactory system. NeuroReport 5:1755–1759
Mendez-Otero R, Santiago MF (2001) Functional role of a glycolipid in directional movements of neurons. An Acad Bras Cienc 73:221–229
Mendez-Otero R, Schlosshauer B, Barnstable CJ, Constantine-Paton M (1988) A developmentally regulated antigen associated with neural cell and process migration. J Neurosci 8:564–579
Mendez-Otero R, Schlosshauer B, Constantine-Paton M (1992) Role of acetylated gangliosides on neuronal migration and axonal outgrowth. In: Lent R (ed) The visual system from genesis to maturity. Birkhauser, Boston, pp 49–62
Messier PE (1978) Microtubules, interkinetic nuclear migration and neurulation. Experientia 34:289–296
Misson JP, Edwards MA, Yamamoto M, Caviness VS Jr (1988) Mitotic cycling of radial glial cells of the fetal murine cerebral wall: a combined autoradiographic and immunohistochemical study. Dev Brain Res 38:183–190
Miyagi T, Wada T, Iwamatsu A, Hata K, Yoshikawa Y, Tokuyama S, Sawada M (1999) Molecular cloning and characterization of a plasma membrane-associated sialidase specific for gangliosides. J Biol Chem 274:5004–5011
Miyakoshi LM, Mendez-Otero R, Hedin-Pereira C (2001a) The 9-O-acetyl GD3 ganglioside is expressed by migrating chains of subventricular zone neurons in vitro. Braz J Med Biol Res 34:669–673
Miyakoshi LM, Mendez-Otero R, Hedin-Pereira C (2001b) Role of 9-O-acetyl GD3 in the migration of subventricular zone neuronal precursors. J Neurochem 78 [Suppl]:56
Morris NR, Efimov VP, Xiang X (1998) Nuclear migration, nucleokinesis and lissencephaly. Trends Cell Biol 8:467–470
Mueller BK (1999) Growth cone guidance: first steps towards a deeper understanding. Annu Rev Neurosci 22:351–358
Multani P, Bonafede DM, Yu RK, Constantine-Paton M (1988) Biochemical characterization of Jones immunoreactive gangliosides in rat. Soc Neurosci Abstr 14:1016
Nadarajah B, Brunstrom JE, Grutzendler J, Wong ROL, Pearlman AL (2001) Two modes of radial migration in early development of the cerebral cortex. Nature Neurosci 4:143–150
Nadarajah B, Alifragis P, Wong ROL, Parnavelas JG (2002) Ventricle-directed migration in the developing cerebral cortex. Nature Neurosci 5:218–224
Noctor SC, Fint AC, Weissman TA, Dammerman RS, Kriegstein AR (2001) Neurons derived from radial glial cells establish radial units in neocortex. Nature 409:714–720
Norgren RB Jr, Lehman MN (1991) Neurons that migrate from the olfactory epithelium in the chick express luteinizing hormone-releasing hormone. Endocrinology 128:1676–1678

Ono K, Kawamura K (1989) Migration of immature neurons along tangentially oriented fibers in the subpial part of the fetal mouse medulla oblongata. Exp Brain Res 78:190–201

Ono M, Handa K, Sonnino S, Withers DA, Nagai H, Hakomori S (2001) GM3 ganglioside inhibits CD9-facilitated haptotactic cell motility: coexpression of GM3 and CD9 is essential in the downregulation of tumor cell motility and malignancy. Biochemistry 40:6414–6421

Pande G (2000) The role of membrane lipids in regulation of integrin function. Curr Opin Cell Biol 12:569–574

Parnavelas JG (2000) The origin and migration of cortical neurons: new vistas. Trends Neurosci 23:126–131

Parnavelas JG, Nadarajah B (2001) Radial glial cells: are they really glia? Neuron 31:881–884

Pearlman AL, Faust PL, Hatten ME, Brunstrom JE (1998) New directions for neuronal migration. Curr Opin Neurobiol 8:45–54

Prinetti A, Marano N, Prioni S, Chigorno V, Mauri L, Casellato R, Tettamanti G, Sonnino S (2000) Association of Src-family protein tyrosine kinase with sphingolipids in rat cerebellar granule cells differentiated in culture. Glycoconj J 17:223–232

Probstmeier R, Pesheva P (1999) Tenascin-C inhibits b1 integrin-dependent cell adhesion and neurite outgrowth on fibronectin by a disialoganglioside-mediated signaling mechanism. Glycobiology 9:101–114

Purpura DP (1978) Ectopic dendritic growth in mature pyramidal neurons in human ganglioside storage disease. Nature 276:520–521

Rakic P (1985) Mechanisms of neuronal migration in developing cerebellar cortex. In: Edelman GE, Cowan WM, Gall E (eds) Molecular basis of neural development. Wiley, New York, pp 139–160

Rakic P (1995) Radial glial cells: scaffolding for brain construction. In: Ketterman H, Ransom BR (eds) Neuroglia. Oxford Univ Press, New York, pp 746–762

Rakic P, Cameron RS, Komuro H (1994) Recognition, adhesion, transmembrane signaling and cell motility in guided neuronal migration. Curr Opin Neurobiol 4:63–69

Raymond AA, Fish DR, Stevens JM, Sisodiya SM, Alsanjari N, Shorvon SD (1994) Subependymal heterotopia: a distinct neuronal migration disorder associated with epilepsy. J Neurol Neurosurg Psychiatry 57:1195–1202

Reinhardt-Maelicke S, Cleeves V, Kindler-Rohrborn A, Rajewsky MF (1990) Differential recognition of a set of O-acetylated gangliosides by monoclonal antibodies RB13-2, D1.1, and Jones during rat brain development. Dev Brain Res 51:279–282

Reitzenstein C, Kopitz J, Schuhmann V, Cantz M (2001) Differential functional relevance of a plasma membrane ganglioside sialidase in cholinergic and adrenergic neuroblastoma cell lines. Eur J Biochem 268:326–333

Renaudin A, Lehmann M, Girault J-A, McKerracher L (1999) Organization of point contacts in neuronal growth cones. J Neurosci Res 55:458–471

Rieber P, Rank G (1994) CDw60: a marker for human CD8+ T helper cells. J Exp Med 179:1385–1390

Ritter G, Boosfeld E, Markstein E, Yu RK, Ren SL, Stallcup WB, Oettgen HF, Old LJ, Livingston PO (1990) Biochemical and serological characteristics of natural 9-O-acetyl GD3 from human melanoma and bovine buttermilk and chemically O-acetylated GD3. Cancer Res 50:1410–1430

Roisen F, Matta SG, Rapport MM (1986) The role of gangliosides in neurotrophic interaction in vitro. In: Tettamati G, Ledeen RW, Sadhoff K, Nagai Y, Toffano G (eds) Gangliosides and neuronal plasticity. Liviana Press, Padova, pp 281–293

Rosner H (1994) Gangliosides and brain development. In: Nicolini M, Zatta PF (eds) Glycobiology and the brain. Pergamon Press, New York, pp 19–39

Ryder EF, Cepko CL (1994) Migration patterns of clonally related granule cells and their progenitors in the developing chick cerebellum. Neuron 12:1011–1029

Santiago MF, Berredo-Pinho M, Costa MR, Gandra M, Cavalcante LA, Mendez-Otero R (2001) Expression and function of ganglioside 9-O-acetyl GD3 in postmitotic granule cell development. Mol Cell Neurosci 17:488–499

Sauer FC (1936) The interkinectic nuclear migration of embryonic epithelial nuclei. J Morphol 60:1–11

Sauer FC, Walker BE (1959) Radiographic study of interkinectic nuclear migration in the neural tube. Proc Soc Exp Biol Med 101:557–560

Schlosshauer B, Blum AS, Mendez-Otero R, Barnstable CJ, Constantine-Paton M (1988) Developmental regulation of ganglioside antigens recognized by the JONES antibody. J Neurosci 8:580–592

Schwarz A, Rapaport E, Hirschberg K, Futerman AH (1995) A regulatory role for sphingolipids in neuronal growth. Inhibition of sphingolipid synthesis and degradation have opposite effects on axonal branching. J Biol Chem 270:10990–10998

Sheikh KA, Sun J, Liu Y, Kawai H, Crawford TO, Proia RL, Griffin JW, Schnaar RL (1999) Mice lacking complex gangliosides develop Wallerian degeneration and myelination defects. Proc Natl Acad Sci USA 96:7532–7537

Shipley MT, McLean JH, Ennis M (1995) Olfactory system. In: Paxinos G (ed) The rat nervous system, 2nd edn. Academic Press, New York, pp 899–922

Sidman R, Rakic P (1982) Development of the human central nervous system. In: Haymaker W, Adams RD (eds) Histology and histopathology of the nervous system. Thomas, Springfield, pp 3–145

Simons M, Friedrichson T, Schultz JB, Pitto M, Masseri M, Kurzchalia TV (1999) Exogenous administration of gangliosides displaces GPI-anchored proteins from lipid microdomains in living cells. Mol Biol Cell 10:3187–3196

Skaper SD, Katoh-Semba R, Varon S (1985) GM1 ganglioside accelerates neurite outgrowth from primary peripheral and central neurons under selected culture conditions. Brain Res 355:19–26

Svennerholm L (1980) Ganglioside designation. In: Svennerholm L, Mandel P, Dreyfus H, Urban PF (eds) Structure and function of gangliosides. Plenum Press, New York, pp 11–13

Takamiya H, Yamamoto A, Furukawa K, Yamashiro S, Shin M, Okada M, Fukumoto S, Haraguchi M, Takeda N, Fujimura K, Sakae M, Kishikawa M, Shiku H, Furukawa K, Aizawa S (1996) Mice with disrupted GM2/GD2 synthase gene lack complex gangliosides but exhibit only subtle defects in their nervous system. Proc Natl Acad Sci USA 93:10662–10667

Takeichi M (1991) Cadherin cell adhesion receptors as a morphogenetic regulator. Science 251:1451–1455

Tamamaki N, Fujimori KE, Takauji R (1997) Origin and route of tangentially migrating neurons in the developing neocortical intermediate zone. J Neurosci 17:8313–8323

Toffano G, Savoini G, Moroni F, Lombardi G, Calza l, Agnati LF (1983) GM1 ganglioside stimulates the regeneration of dopaminergic neurons in the central nervous system. Brain Res 261:163–166

Uher BF, Golden JA (2000) Neuronal migration defects of the cerebral cortex: a destination debacle. Clin Genet 58:16–24

Varki A, Hooshmand F, Diaz S, Varki NM, Hedrick SM (1991) Developmental abnormalities in transgenic mice expressing a sialic acid-specific 9-O-acetylase. Cell 65:65–74

Vyas AA, Schnaar RL (2001) Brain gangliosides: functional ligands for myelin stability and the control of nerve regeneration. Biochimie 83:677–682

Walsh CA, Goffinet AM (2000) Potential mechanisms of mutations that affect neuronal migration in man and mouse. Curr Opin Genet Dev 10:270–274

Walsh FS, Doherty P (1997) Neural cell adhesion molecules of the immunoglobulin superfamily: role in axon growth and guidance. Annu Rev Cell Biol 13:425–456

Wichterle H, Garcia-Verdugo J, Alvarez-Buylla A (1997) Direct evidence for homotypic, glia-independent neuronal migration. Neuron 18:779–791

Yamashita T, Wada R, Sasaki T, Deng C, Bierfreund U, Sandhoff K, Proia RL (1999) A vital role for glycosphingolipid synthesis during development and differentiation. Proc Natl Acad Sci USA 96:9142–9147

Yang LJ-S, Zeller CB, Shaper NL, Kiso M, Hasegawa A, Shapiro RE, Schnaar RL (1996) Gangliosides are neuronal ligands for myelin-associated glycoprotein. Proc Natl Acad Sci USA 93:814–818

Yates AJ, Rampersaud A (1998) Sphingolipids as receptor modulators: an overview. Ann NY Acad Sci 845:7–71

Yee KT, Simon HH, Tessier-Lavigne M, O'Leary DM (1999) Extension of long leading processes and neuronal migration in the mammalian brain directed by the chemoattractant netrin-1. Neuron 24:607–622

Yu RK (1994) Developmental regulation of ganglioside metabolism. Prog Brain Res 10:3–44

Myelin Molecules Limiting Nervous System Plasticity

R.L. Schnaar

1 Introduction

During development, nerve growth cones are guided by a balance of positive and negative cues that ensure robust axon outgrowth and fine-tuned control of direction and trajectory (Stoeckli and Landmesser 1998). In contrast to the embryonic environment, the injured adult central nervous system (CNS) in mammals is overwhelmingly inhibitory for axon outgrowth, severely limiting nerve regeneration (Fry 2001). The inability of axons to regrow after CNS injury has profound medical consequences. This is most evident in the lack of recovery from spinal cord injuries, which result in life-long loss of function for millions of people worldwide (Geisler et al. 2001; Sekhon and Fehlings 2001). The pathology of brain trauma, stroke, and progressive multiple sclerosis are also negatively impacted by limitations on axon outgrowth in the CNS. The discovery of the molecular mechanisms that render the CNS such an inhospitable environment for nerve regeneration may provide new approaches to enhance recovery of nerve damage after injury or disease (Kwon and Tetzlaff 2001).

Axons fail to extend more than a few millimeters beyond a CNS injury site, even after carefully controlled "microcrush" experimental lesions (Selles-Navarro et al. 2001). However, the nerves indeed have the intrinsic ability to regenerate given the appropriate environment. This was experimentally demonstrated by taking advantage of the difference in regeneration in the CNS and the peripheral nervous system (PNS). In contrast to the CNS, when axons are damaged in the PNS they often successfully regrow over long distances through the distal nerve sheath. When David and Aguayo (1981) transplanted a segment of peripheral nerve into injured CNS tissue, they found that the adjacent CNS axons

Departments of Pharmacology and Neuroscience, The Johns Hopkins School of Medicine, 725 N. Wolfe Street, Baltimore, Maryland 21205, USA

Progress in Molecular and Subcellular Biology, Vol. 32
Ivica Kostović (Ed.)

readily regenerated through the transplanted PNS environment. This and subsequent experiments led to the principle that endogenous factors in the local environment actively inhibit nerve regeneration, despite the nerve's intrinsic ability to regenerate. In this light, the distal end of the injured axon is exposed to distinct sets of molecular inhibitors in the glial scar, which forms at the injury site, and on myelin, which persists at and distal to the injury site (Lehmann et al. 1999; Brittis and Flanagan 2001; Fournier and Strittmatter 2001; Schwab 2002).

Immediately after an injury to the CNS, inflammatory processes initiate a rapid cascade of secondary damage resulting in progressive cavitation and formation of an astrocytic glial scar (Fitch et al. 1999). This process is responsible for a significant amount of the acute tissue destruction, and strategies to minimize the cellular and biochemical processes responsible may provide the best short-term opportunities to lessen the severity of CNS injuries. The glial scar which forms at the injury site contains molecules which persist and inhibit axon outgrowth (McKeon et al. 1991, 1999; Becker et al. 2000), including chondroitin sulfate proteoglycans and tenascin-R. In addition to the glial scar, residual CNS myelin contains molecules that inhibit axon outgrowth. The relative importance of the glial scar and myelin inhibitors in limiting nerve regeneration remains a matter of experimental study.

Three types of CNS myelin molecules inhibit nerve regeneration (Table 1): Nogo, myelin-associated glycoprotein (MAG), and chondroitin sulfate proteoglycans (CSPGs; Fournier and Strittmatter 2001). Each was identified by its ability to inhibit neurite outgrowth in vitro. Recent studies provide significant insights about these inhibitors. Even so, our knowledge remains rudimentary, and the mechanisms by which each regulates axon outgrowth, and the relative role of each in limiting nerve regeneration in vivo are matters yet to be resolved.

Since independent data support roles for Nogo, MAG and CSPGs in limiting nerve regeneration, it is prudent to investigate each of these in detail in the hope that the resulting knowledge can be integrated into enhanced understanding of why functional recovery from CNS injuries is so profoundly limited.

2 Nogo

Subsequent to the studies of Aguayo and coworkers, demonstrating that CNS nerves regenerate through transplanted PNS nerve sheathes in vivo (David and Aguayo 1981; Benfey and Aguayo 1982), Schwab and

Table 1. Myelin inhibitors of nerve regeneration

Inhibitor	Molecular characteristics	Cell and tissue distribution	Nerve target(s)	Second messengers	Primary references
Nogo			NgR	Unknown	Chen et al. (2000),
Nogo A	130-kDa protein	Oligodendrocytes, some CNS neurons, others?			GrandPre et al. (2000), Fournier et al. (2001)
Nogo B	40-kDa protein	CNS, PNS, kidney, lung, others?			
Nogo C	22-kDa protein	CNS, PNS, muscle, others?			
MAG[a]	100-kDa glycoprotein	Oligodendrocytes and Schwann cells	Gangliosides, GD1a and GT1b, glycoproteins (?)	RhoA, protein kinase A	McKerracher et al. (1994), Mukhopadhyay et al. (1994), Lehmann et al. (1999), Cai et al. (1999), Vyas et al. (2002)
CSPGs[b]	Proteoglycans	Glial scar, myelin, broadly distributed at cell surface and extracellular matrix	Unknown	Unknown	Snow et al. (1990), McKeon et al. (1991), Niederost et al. (1999), Bradbury et al. (2002)

[a]Myelin-associated glycoprotein.
[b]Chondroitin-sulfate proteoglycans.

Thoenen (1985) replicated this finding in an in vitro system in which ganglion cells were found to regenerate through sciatic, but not optic, nerve bridges. Subsequently, Caroni and Schwab demonstrated that PNS myelin adsorbed to a culture substratum supported neurite outgrowth, whereas CNS myelin failed to do so (Caroni and Schwab 1988b; Schwab and Caroni 1988). The inhibitory factors on CNS myelin were protease-sensitive, and two fractions isolated after SDS-PAGE resolution of rat CNS myelin proteins, later designated NI-250 and NI-35, contained significant neurite outgrowth inhibitory activity (Caroni and Schwab 1988b). These fractions also inhibited fibroblast spreading, providing a facile functional assay. A monoclonal antibody cross-reactive with NI-35/250, designated IN-1, neutralized both their neurite outgrowth and fibroblast spreading inhibitory activities (Caroni and Schwab 1988a). Using IN-1 binding and the above bioassays, the bovine form of NI-250 (bNI-220) was purified by SDS-PAGE, and the peptide sequence was obtained (Spillmann et al. 1998). With the peptide sequence in hand, three groups cloned the gene from rat and human cDNA libraries (Chen et al. 2000; GrandPre et al. 2000; Prinjha et al. 2000). The resulting gene was designated *nogo* and the three polypeptides arising from the gene by alternative promoter use and splicing were designated Nogo-A (130 kDa), -B (40 kDa), and -C (22 kDa).

Human Nogo-A and -B share identical N-terminal (183 amino acids) and C-terminal (188 amino acids) domains, with Nogo-A having an additional ~800 amino acids separating the two. Nogo-C consists of the same 188-amino acid C-terminal domain with a unique 11 amino acid N-terminal extension. Rat and human share over 80% identity in the N-terminal domain, and a remarkable 97% identity in the shared C-terminal 188 amino acids. Most of Nogo is found in the endoplasmic reticulum (ER), with a small amount appearing at the cell surface. Although there is no signal sequence, there are two putative transmembrane domains in the highly conserved C-terminal domain that are separated by 66 amino acids. Although alternate orientations have been posited (Huber and Schwab 2000), convincing data support a model in which the N- and C-termini are cytoplasmically oriented with the intervening 66-amino acid stretch (Nogo-66) facing the ER lumen and the extracellular space, where it may interact with neurons to inhibit nerve regeneration (Fournier et al. 2001; GrandPre et al. 2000).

In situ hybridization studies indicate that Nogo mRNA is expressed by oligodendrocytes (but not by Schwann cells) and by a subset of neurons in the CNS (Chen et al. 2000; GrandPre et al. 2000). An antibody against Nogo-66 reveals surface staining on the cell surface of isolated chick oligodendrocytes in cell culture (GrandPre et al. 2000). Northern analyses indicate that Nogo is not restricted to the CNS, although it does not appear to be expressed in the PNS (Chen et al. 2000; GrandPre et al. 2000). Since most of Nogo is oriented in the ER lumen (it has an ER retention signal and is related to the reticulon family of ER proteins), it is predicted to have intracellular function(s) separate from its axon outgrowth inhibitory role(s), although these functions have yet to be elucidated.

Nogo-A has two inhibitory polypeptide domains (Fournier et al. 2001): (1) the highly acidic N-terminal domain has a lower-affinity inhibitor which is not cell type-specific, in that it inhibits fibroblast spreading as well as neurite outgrowth in vitro, and (2) Nogo-66, the short predicted extracellular domain, is a high affinity inhibitor specific for neuronal growth cone collapse. The large N-terminal domain, which may be facing the cytoplasm (GrandPre et al. 2000), might be expected to become accessible to the nerve cell surface only after myelin or oligodendrocyte damage. Nevertheless, an antibody raised to a peptide in the middle of the N-terminal domain enhanced neurite outgrowth through optic nerve bridges in vitro (Chen et al. 2000), and chimeric molecules lacking the C-terminal domain inhibit neurite outgrowth and fibroblast spreading in vitro (Prinjha et al. 2000; Fournier et al. 2001), although they appear to require either immobilization or polyvalent cross-linking, suggesting a low site affinity multivalent mechanism of action. The antibody IN-1 reverses the inhibition of both axon outgrowth and fibroblast spreading, suggesting it is against the N-terminal inhibitory domain.

In contrast to the N-terminal Nogo-A domain, Nogo-66 specifically inhibits axon outgrowth with nanomolar affinity even when monovalent (Fournier et al. 2001). Using a chimeric molecule with Nogo-66 linked to the reporter alkaline phosphatase, Fournier et al. (2001) expression cloned a receptor for Nogo-66, NgR, from a mouse brain cDNA library. NgR is a glycophosphatidylinositol-linked (GPI-linked) protein with a peptide mass of 51 kDa that is 90% identical between mouse and human. It has eight leucine-rich repeat sequences which place it in a large "LLR" family, some of which are involved in cell-cell and cell-matrix interactions (Hocking et al. 1998; Kobe and Deisenhofer 1994), including the slit family of growth cone regulatory molecules (Van Vactor and Flanagan 1999). Northern analysis and in situ hybridization demonstrated that NgR mRNA is expressed predominantly in brain (but also in heart and kidney) and appears in selected neurons in gray matter, but is not detected in white matter. NgR was immunolocalized on the surface of axons of chick spinal neurons in vitro.

Nogo sensitivity is absent from neurons during early development, and is gained near the time of myelination (Bandtlow and Loschinger 1997). Likewise, NgR expression is low or absent in chick dorsal root ganglion (DRG) neurons at 7 days of embryonic development (E7), but is apparent in E13 DRG neurons. Consistent with this finding, E12 DRG neurons are sensitive to Nogo-66, whereas E7 DRG neurons are not, unless first infected with a viral construct that induces NgR expression (Fournier et al. 2001). Furthermore, treatment of E12 DRG neurons with an enzyme that specifically cleaves GPI-anchored proteins from the membrane renders them insensitive to Nogo-66. Together, these data strongly implicate Nogo-66 and its GPI-linked receptor, NgR, as key molecular determinants limiting axon outgrowth in the adult.

Functional studies on Nogo in vivo have focused on use of the monoclonal antibody IN-1 to enhance recovery after CNS injury. Since IN-1 reverses Nogo-mediated inhibition of both axon outgrowth and fibroblast spreading, it likely binds to the N-terminal domain of Nogo. Studies using CNS-implanted hybridomas secreting IN-1 antibody revealed enhanced recovery of function in rats subjected to dorsal hemisection of the spinal cord at the T8 level (Merkler et al. 2001), or corticospinal tract lesions at the level of the medulla oblongata (Raineteau et al. 2001). Anatomical studies have demonstrated both enhanced axon regeneration and enhanced collateral innervation in response to IN-1 antibody. A humanized recombinant IN-1 perfused onto the site of a thoracic spinal cord dorsolateral hemisection in rats also resulted in some axon extension beyond the injury site (Brosamle et al. 2000). These results are encouraging, although their significance for the treatment of nerve injury in humans has yet to be evaluated. Therapeutics targeting the Nogo-66 high affinity inhibitory domain may have enhanced anti-Nogo effects; experiments evaluating this possibility have not yet been reported.

3
Myelin-Associated Glycoprotein

Myelin-associated glycoprotein (MAG) is a quantitatively minor constituent of CNS myelin, where it comprises ~1% of myelin protein (Trapp 1990). It is also found in PNS, although at markedly lower concentrations (~0.1% of myelin protein). In the adult CNS, MAG expression is highly restricted to the innermost (periaxonal) wrap of membrane directly across from the axon (Trapp et al. 1989), where it has been posited to interact with complementary binding molecules on the apposing axon surface. In the PNS, MAG is more broadly distributed, appearing on periaxonal membranes, but also on paranodal loops, Schmidt-Lanterman incisures, and mesaxon membranes (Trapp et al. 1989).

MAG-deficient mice myelinate axons, but display age-dependent progressive axon atrophy, axon degeneration and demyelination (Fruttiger et al. 1995; Yin et al. 1998; Weiss et al. 2001), as well as delayed myelination of optic nerves (Bartsch et al. 1997). The axonal atrophy has been associated with decreased neurofilament phosphorylation and spacing (Yin et al. 1998). These data support a model in which MAG binds to complementary receptor molecules on the axon surface resulting in transmembrane signaling which enhances axon cytoskeletal structure and stabilizes the axon-myelin interaction. In addition to this essential function in axon-myelin stabilization, MAG also serves as an inhibitor of nerve regeneration.

When proteins were solubilized from myelin membranes using a mild detergent, then resolved into separate species by ion exchange chromatography, some of the fractions inhibited neurite outgrowth from a neural cell line in vitro (McKerracher et al. 1994). One peak of inhibitory activity coincided with the elution of MAG, leading the investigators to directly test a recombinant form of MAG. This resulted in clear evidence that MAG was an inhibitory factor extracted from myelin. In separate experiments, neurite outgrowth was inhibited when primary cerebellar granule neurons isolated from neonatal rats were seeded onto monolayers of fibroblasts engineered to stably express MAG (Mukhopadhyay et al. 1994). Together, these findings support a role for MAG as an inhibitor of nerve regeneration. Subsequent in vitro studies indicated that the neurite inhibitory effect of MAG was observed using a variety of neuronal cell types (DeBellard et al. 1996). As with Nogo (Bandtlow and Loschinger 1997), sensitivity of some neurons to MAG inhibition is developmentally regulated (DeBellard et al. 1996).

MAG binds to the surface of neurons (Poltorak et al. 1987) and initiates a signal that results in growth cone collapse and inhibition of axon outgrowth (Li et al. 1996). Important insight into the target ligands on the nerve cell surface which bind to MAG came when MAG was identified as a member of the "Siglec" family (Kelm et al. 1994; Crocker et al. 1998; Crocker and Varki 2001). Siglecs are "sialic acid binding immunoglobulin-family member lectins," denoting that they have a polypeptide domain designed to bind to the sugar "sialic acid". Sialic acids are major glycan determinants on the surface of all vertebrate cells, including neurons. The implication is that MAG binds to nerve cell surface sialic acid-containing structures (glycoproteins or glycolipids) to initiate its biological effects. This prediction has been borne out in vitro; treatment of neurons with neuraminidase, which enzymatically removes cell surface sialic acids, attenuated MAG binding and MAG-mediated effects on neurite outgrowth (DeBellard et al. 1996; Vyas et al. 2002).

The brain is particularly enriched in a family of sialoglycoconjugates termed gangliosides, which are glycolipids consisting of a ceramide lipid inserted in the cell membrane and a glycan, which extends into the extracellular space (Schnaar 2000). The four major gangliosides found in the brain, GM1, GD1a, GD1b, and GT1b (Schnaar 2000; Rösner, this Vol.), are common to all mammals and birds (Tettamanti et al. 1973). Several lines of evidence implicate gangliosides GD1a and GT1b as functional MAG ligands. Initial binding studies demonstrated that MAG bound preferentially to the terminal sugar sequence "NeuAc α3 Gal β3 GalNAc," which is also termed the "3–O" structure (Crocker et al. 1996). The two major brain gangliosides are the predominant molecules that carry this structure on nerve surfaces. This led to studies that demonstrated that GD1a and GT1b selectively support MAG binding (Yang et al. 1996; Collins et al. 1997). However, binding, per se, does not demonstrate function (MAG and other Siglecs bind to undetermined sialic acids on red blood cells, although there is no function related to this binding; Kelm et al. 1994). To test the function of gangliosides, more powerful approaches were required. One such approach used mice in which complex gangliosides were genetically deleted (Liu et al. 1999).

Biosynthesis of gangliosides progresses in a stepwise manner, each sugar being added by a specific glycosyltransferase (Schnaar 2000; Rösner, this Vol.). Genetic disruption of the gene for one enzyme in this pathway (Liu et al. 1999), an N-acetylgalactosaminyltransferase, resulted in mice which no longer expressed complex gangliosides (such as GD1a and GT1b), but instead expressed a comparable concentration of the simpler gangliosides GD3, which is not a MAG ligand, and GM3, which is a relatively poor MAG ligand (Collins et al. 1997). These mice displayed an axon degeneration and dysmyelination phenotype similar to that found in mice lacking MAG (Sheikh et al. 1999), supporting the hypothesis that GD1a and GT1b are functional MAG ligands. Furthermore, they displayed selective and progressive downregulation of MAG expression (Sheikh et al. 1999; Sun and Schnaar 2000), implicating MAG-ganglioside binding in the long-term steady-state expression of MAG. The functional consequences of MAG-ganglioside binding were evidenced by progressive motor behavioral deficits in mice with a disrupted ganglioside N-acetylgalactosaminyltransferase gene (Chiavegatto et al. 2000). These data implicate gangliosides in MAG-mediated stability of myelin-axon interactions. Additional studies implicated MAG-ganglioside binding in the inhibition of nerve regeneration (Vyas et al. 2002).

MAG-mediated neurite outgrowth inhibition in vitro was attenuated by: (1) neuraminidase treatment of the neurons; (2) blocking neuronal ganglioside biosynthesis using a selective inhibitor of glycosphingolipid biosynthesis (1-phenyl-2-hexadecanoylamino-3-morpholino-1-propanol); (3) using nerve cells from ganglioside N-acetylgalactosaminyltransferase-knockout mice; and (4) adding highly specific IgG-class anti-GD1a and anti-GT1b monoclonal antibodies (Vyas et al. 2002). These data implicate the gangliosides GD1a and GT1b as functional MAG ligands in the inhibition of nerve regeneration. Although these data implicate gangliosides as functional MAG ligands, they do not exclude other potential ligands. Some published data identify proteins that bind to MAG (De Bellard and Filbin 1999; Strenge et al. 2001), although their functions remain to be established.

Although gangliosides are implicated as functional MAG ligands, the mechanism by which MAG-ganglioside binding is translated into changes in nerve cell structure and function have yet to be established. An initial step in MAG-nerve cell signaling may be ganglioside cross-linking at the nerve cell surface. Addition of a naturally multivalent IgM-class anti-GT1b antibody to rat embryonic hippocampal neurons resulted in inhibition of neurite outgrowth (Vinson et al. 2001). By comparison, divalent IgG-class anti-ganglioside antibodies to GD1a or GT1b have no effect on neurite outgrowth in the absence of MAG. However, pre-clustering the same antibodies with a secondary antibody generates a reagent that, when added to rat cerebellar granule cells in vitro, strongly inhibits neurite outgrowth (Vyas et al. 2002). Thus, ganglioside clustering appears to be a first step in MAG-mediated inhibition of nerve regeneration. Subsequent steps in the signal transduction cascade are discussed in a following section.

The relative role of MAG compared to other inhibitors in the inhibition of nerve regeneration after injury in vivo remains to be established. In one model (Schäfer et al. 1996), mice genetically engineered to lack MAG were crossed with mice having delayed peripheral nerve Wallerian degeneration (*Wld^s*). Peripheral nerve regeneration is suppressed along persisting myelin sheathes in Wld^s mice. However, animals that did not express MAG demonstrated a twofold increase in nerve regeneration along myelin sheathes in Wld^s mice compared to animals that expressed MAG. These data implicate a role for MAG inhibition in restricting nerve regeneration in vivo.

4 Chondroitin Sulfate Proteoglycans

Chondroitin sulfate proteoglycans (CSPGs) are proteins post-translationally modified with glycosaminoglycan chains consisting of long repeating N-acetylgalactosamine-glucuronic acid residues derivatized with sulfate esters (Bovolenta and Fernaud-Espinosa 2000). CSPGs are widely distributed as cell surface and extracellular matrix molecules. Evidence for their role in inhibiting nerve regeneration arose from the findings that (1) CSPG expression is upregulated at the sites of CNS injury (McKeon et al. 1991); and (2) CSPGs are inhibitory for neurite outgrowth in vitro (Snow et al. 1990). CSPGs are present both in the astroglial scar (Asher et al. 2001) and on myelin (Niederost et al. 1999), and may contribute to the inhibitory nature of both. Enzymatic degradation of the chondroitin sulfate saccharide chains reverses the inhibitory effects of CSPGs on neurite outgrowth (Snow et al. 1990), indicating that the sugar chains, per se, are required for the inhibitory function of CSPGs. Whether the protein portion also plays a role is unclear. Several CSPGs (neurocan, versican, NG2, phosphocan, and others) are upregulated at sites of nerve injury (Asher et al. 2001). However, evidence is accumulating that one, NG2, which is expressed on oligodendrocyte precursors, macrophages, and some reactive astrocytes, may be a particularly important inhibitor after CNS injury (Fidler et al. 1999; Jones et al. 2002). The structural specificity of CSPGs for inhibition of nerve regeneration has not been fully established. However, at sites of experimental injury nerves regenerate through regions rich in heparan sulfate proteoglycans (which are closely related in structure to CSPGs), but not through nearby regions rich in CSPGs (Moon et al. 2002). Whether this is due to intrinsic differences in the ability of the two types of sugar chains to inhibit nerve regeneration, or due to other molecules in these regions has yet to be established. Studies using bacterial chondroitinase ABC, an enzyme that degrades CSPG glycosaminoglycan chains, demonstrate enhanced nerve regeneration and enhanced function in an animal model of spinal cord injury (Bradbury et al. 2002). Complementary nerve cell surface receptors for CSPGs, through which inhibitory signals may be mediated, have yet to be identified.

5 Myelin Vaccine

As detailed above, there are multiple identified (and perhaps additional unidentified) inhibitors of nerve regeneration expressed on myelin. One challenge is to identify which of these are most relevant in limiting nerve regeneration after injury. The use of genetically engineered (knockout) mice, enzymes, and antibodies to neutralize one or more of the inhibitors in animal models of nerve injury will likely help to resolve this issue. Therapies designed to enhance nerve regeneration may have to block multiple myelin inhibitors. One experimental approach to achieve this goal involved immunizing mice with purified myelin, which presumably contains all relevant inhibitors (Huang et al. 1999). Myelin-immunized mice that were subsequently subjected to experimental dorsal hemisection of the spinal cord demonstrated improved axon regeneration and improvement of function without overt signs of an anti-myelin inflammatory response. These data are encouraging, in that they indicate that blocking myelin inhibitors in combination may enhance nerve regeneration.

6
Signaling Pathways in the Inhibition of Nerve Regeneration

Presumably, each myelin inhibitor of nerve regeneration binds to a specific complementary molecule on the surface of the nerve cell, initiating signal transduction cascades that result in growth cone collapse and inhibition of axon extension. An understanding of the network of signaling pathways that lead from inhibitor binding to axon outgrowth inhibition may provide additional opportunities for therapeutic intervention. Although knowledge in this area is limited, two classes of intracellular messengers have thus far been implicated: cyclic AMP and RhoA.

Cyclic nucleotide involvement in neurite outgrowth inhibition was demonstrated in spinal neurons isolated from *Xenopus* embryos (Song et al. 1998). Growth cones from these neurons turn away from a micropipette filled with a soluble recombinant form of MAG. In contrast, after activation of protein kinase A in these same cells (using a cAMP analog, adenosine-3′,5′ – cyclic monophosphorothioate, Sp-isomer) the response to MAG was reversed from repulsion to attraction. Subsequent experiments using rat cerebellar neurons extended these findings, in that pre-treatment of cells with cAMP or agents that increased cAMP (including neurotrophins) attenuated the inhibitory effects of MAG (Cai et al. 1999). Although these experiments place the activation of protein kinase A in a signaling network that controls the growth cone response to MAG, they do not implicate cAMP as a direct second messenger in the inhibitory response to MAG. Instead, myelin-mediated inhibition of nerve regeneration is likely to involve a network of interacting signal transduction pathways, with cAMP implicated as a response regulator.

Members of the Rho family of small GTPases mediate signal transduction from extracellular signals to the actin cytoskeleton in eukaryotic cells from yeast to humans (Luo 2000), and are key regulators of growth cone dynamics. Members of this family of small intracellular proteins act as molecular switches, each having a GTP-bound "on" configuration and a GDP-bound "off" configuration. In turn, each Rho family member is activated by guanine nucleotide exchange factors (GEFs) that promote the exchange of bound GDP for GTP, and are inactivated by GTPase-activating proteins (GAPs) that promote the intrinsic enzymatic cleavage of bound GTP to GDP. There are three major Rho family members, RhoA, Rac and Cdc42, each of which is linked to extracellular signals by multiple GEFs and GAPs. Rac and Cdc42 are generally positive regulators for growth cone extension, whereas RhoA is generally a negative regulator, involved in growth cone collapse or arrest.

There is clear evidence for RhoA involvement downstream of inhibitors of nerve regeneration in vitro and in vivo (Lehmann et al. 1999). MAG- or myelin-coated substrata inhibit neurite outgrowth from a neuronal cell line, PC12. However, MAG inhibition was completely reversed when the cells were pre-treated with C3, a specific inhibitor of RhoA (C3 is a Clostridial enzyme which specifically ADP-ribosylates RhoA). Consistent with this finding, PC12 cells stably transfected with a dominant negative form of RhoA were completely resistant to MAG-mediated inhibition of neurite outgrowth (although they remained susceptible to inhibition by myelin). The relevance of these findings to CNS injury was tested in an animal model consisting of a microcrush lesion of the optic nerve (Selles-Navarro et al. 2001). In untreated animals, lesioned axons do not regenerate beyond the crush site. However, when the lesioned nerve was exposed to C3, many axons progressed beyond the microcrush site and into the distal area that contained residual myelin.

The data described above suggest that inhibiting RhoA and/or activating protein kinase A may be therapeutic strategies to enhance nerve regeneration after injury. However, both of these are ubiquitous and potent signaling molecules, and their manipulation will likely result in unacceptable side effects unless treatments are delivered in a narrowly targeted manner. Alternatively, as specific upstream and downstream factors are identified in the signaling pathways controlling the inhibition of nerve regeneration, more specific targets may emerge.

7 Concluding Remarks

During development, the nervous system is highly plastic, with nerves extending axons and making synapses in excess of what is required, later to be pared appropriately. As the mammalian CNS ages, plasticity is sharply reduced, until the environment is strongly inhibitory for axon outgrowth. The reason for this shift has not been established, and remains a matter for speculation (Goldberg and Barres 2000). It may be that in the densely packed milieu of the CNS, promiscuous connectivity impairs function, so the expression of inhibitors is evolutionarily favored. This is sensible, in that myelination occurs late in nervous system development, just as CNS plasticity decreases, so evolutionary selection of myelin as a source of inhibitors of axon outgrowth would be developmentally timely. As animal models lacking the specific myelin inhibitors of axon outgrowth are developed, theories concerning the normal function of inhibition in the adult CNS can be addressed.

Regardless of their physiological roles, myelin inhibitors profoundly restrict recovery from CNS injury. As far back as 4000 years ago, a surgical papyrus was reported to have instructed, "One having a crushed vertebra in his neck; he is unconscious of his two arms, his two legs, he is speechless. An ailment not to be treated" (Brittis and Flanagan 2001). The long-held notion that adult CNS injury is typically beyond cure still holds today. However, a molecular understanding of some of the impediments to nerve regeneration is emerging, and this knowledge may eventually provide new therapeutic approaches to enhance functional recovery after CNS injury.

Acknowledgements. The author is grateful to the National Institutes of Health, the National Multiple Sclerosis Society, and the Stollof Family Fund for financial support, and to Dr. Alka A. Vyas for editorial considerations.

References

Asher RA, Morgenstern DA, Moon LD, Fawcett JW (2001) Chondroitin sulphate proteoglycans: inhibitory components of the glial scar. Prog Brain Res 132:611–619

Bandtlow CE, Loschinger J (1997) Developmental changes in neuronal responsiveness to the CNS myelin-associated neurite growth inhibitor NI-35/250. Eur J Neurosci 9:2743–2752

Bartsch S, Montag D, Schachner M, Bartsch U (1997) Increased number of unmyelinated axons in optic nerves of adult mice deficient in the myelin-associated glycoprotein (MAG). Brain Res 762:231–234

Becker T, Anliker B, Becker CG, Taylor J, Schachner M, Meyer RL, Bartsch U (2000) Tenascin-R inhibits regrowth of optic fibers in vitro and persists in the optic nerve of mice after injury. Glia 29:330–346

Benfey M, Aguayo AJ (1982) Extensive elongation of axons from rat brain into peripheral nerve grafts. Nature 296:150–152

Bovolenta P, Fernaud-Espinosa I (2000) Nervous system proteoglycans as modulators of neurite outgrowth. Prog Neurobiol 61:113–132

Bradbury EJ, Moon LD, Popat RJ, King VR, Bennett GS, Patel PN, Fawcett JW, McMahon SB (2002) Chondroitinase ABC promotes functional recovery after spinal cord injury. Nature 416:636–640

Brittis PA, Flanagan JG (2001) Nogo domains and a Nogo receptor: implications for axon regeneration. Neuron 30:11–14

Brosamle C, Huber AB, Fiedler M, Skerra A, Schwab ME (2000) Regeneration of lesioned corticospinal tract fibers in the adult rat induced by a recombinant, humanized IN-1 antibody fragment. J Neurosci 20:8061–8068

Cai D, Shen Y, De Bellard M, Tang S, Filbin MT (1999) Prior exposure to neurotrophins blocks inhibition of axonal regeneration by MAG and myelin via a cAMP-dependent mechanism. Neuron 22:89–101

Caroni P, Schwab ME (1988a) Antibody against myelin-associated inhibitor of neurite growth neutralizes non-permissive substrate properties of CNS white matter. Neuron 1:85–96

Caroni P, Schwab ME (1988b) Two membrane protein fractions from rat central myelin with inhibitory properties for neurite growth and fibroblast spreading. J Cell Biol 106:1281–1288

Chen MS, Huber AB, van der Haar ME, Frank M, Schnell L, Spillmann AA, Christ F, Schwab ME (2000) Nogo-A is a myelin-associated neurite outgrowth inhibitor and an antigen for monoclonal antibody IN-1. Nature 403:434–439

Chiavegatto S, Sun J, Nelson RJ, Schnaar RL (2000) A functional role for complex gangliosides: motor deficits in GM2/GD2 synthase knockout mice. Exp Neurol 166:227–234

Collins BE, Kiso M, Hasegawa A, Tropak MB, Roder JC, Crocker PR, Schnaar RL (1997) Binding specificities of the sialoadhesin family of I-type lectins. Sialic acid linkage and substructure requirements for binding of myelin-associated glycoprotein, Schwann cell myelin protein, and sialoadhesin. J Biol Chem 272:16889–16895

Crocker PR, Varki A (2001) Siglecs in the immune system. Immunology 103:137–145

Crocker PR, Kelm S, Hartnell A, Freeman S, Nath D, Vinson M, Mucklow S (1996) Sialoadhesin and related cellular recognition molecules of the immunoglobulin superfamily. Biochem Soc Trans 24:150–156

Crocker PR, Clark EA, Filbin M, Gordon S, Jones Y, Kehrl JH, Kelm S, Le Douarin N, Powell L, Roder J, Schnaar RL, Sgroi DC, Stamenkovic K, Schauer R, Schachner M, van den Berg TK, van der Merwe PA, Watt SM, Varki A (1998) Siglecs: a family of sialic-acid binding lectins. Glycobiology 8(2):v

David S, Aguayo AJ (1981) Axonal elongation into peripheral nervous system "bridges" after central nervous system injury in adult rats. Science 214:931–933

De Bellard ME, Filbin MT (1999) Myelin-associated glycoprotein, MAG, selectively binds several neuronal proteins. J Neurosci Res 56:213–218

DeBellard ME, Tang S, Mukhopadhyay G, Shen Y-J, Filbin MT (1996) Myelin-associated glycoprotein inhibits axonal regeneration from a variety of neurons via interaction with a sialoglycoprotein. Mol Cell Neurosci 7:89–101

Fidler PS, Schuette K, Asher RA, Dobbertin A, Thornton SR, Calle-Patino Y, Muir E, Levine JM, Geller HM, Rogers JH, Faissner A, Fawcett JW (1999) Comparing astrocytic cell lines that are inhibitory or permissive for axon growth: the major axon-inhibitory proteoglycan is NG2. J Neurosci 19:8778–8788

Fitch MT, Doller C, Combs CK, Landreth GE, Silver J (1999) Cellular and molecular mechanisms of glial scarring and progressive cavitation: in vivo and in vitro analysis of inflammation-induced secondary injury after CNS trauma. J Neurosci 19:8182–8198

Fournier AE, Strittmatter SM (2001) Repulsive factors and axon regeneration in the CNS. Curr Opin Neurobiol 11:89–94

Fournier AE, GrandPre T, Strittmatter SM (2001) Identification of a receptor mediating Nogo-66 inhibition of axonal regeneration. Nature 409:341–346

Fruttiger M, Montag D, Schachner M, Martini R (1995) Crucial role for the myelin-associated glycoprotein in the maintenance of axon-myelin integrity. Eur J Neurosci 7:511–515

Fry EJ (2001) Central nervous system regeneration: mission impossible? Clin Exp Pharmacol Physiol 28:253–258

Geisler FH, Coleman WP, Grieco G, Poonian D (2001) Measurements and recovery patterns in a multicenter study of acute spinal cord injury. Spine 26:S68–S86

Goldberg JL, Barres BA (2000) Nogo in nerve regeneration. Nature 403:369–370

GrandPre T, Nakamura F, Vartanian T, Strittmatter SM (2000) Identification of the Nogo inhibitor of axon regeneration as a Reticulon protein. Nature 403:439–444

Hocking AM, Shinomura T, McQuillan DJ (1998) Leucine-rich repeat glycoproteins of the extracellular matrix. Matrix Biol 17:1–19

Huang DW, McKerracher L, Braun PE, David S (1999) A therapeutic vaccine approach to stimulate axon regeneration in the adult mammalian spinal cord. Neuron 24:639–647

Huber AB, Schwab ME (2000) Nogo-A, a potent inhibitor of neurite outgrowth and regeneration. Biol Chem 381:407–419

Jones LL, Yamaguchi Y, Stallcup WB, Tuszynski MH (2002) NG2 is a major chondroitin sulfate proteoglycan produced after spinal cord injury and is expressed by macrophages and oligodendrocyte progenitors. J Neurosci 22:2792–2803

Kelm S, Pelz A, Schauer R, Filbin MT, Song T, de Bellard ME, Schnaar RL, Mahoney JA, Hartnell A, Bradfield P, Crocker PR (1994) Sialoadhesin, myelin-associated glycoprotein and CD22 define a new family of sialic acid-dependent adhesion molecules of the immunoglobulin superfamily. Curr Biol 4:965–972

Kobe B, Deisenhofer J (1994) The leucine-rich repeat: a versatile binding motif. Trends Biochem Sci 19:415–421

Kwon BK, Tetzlaff W (2001) Spinal cord regeneration: from gene to transplants. Spine 26:S13–S22

Lehmann M, Fournier A, Selles-Navarro I, Dergham P, Sebok A, Leclerc N, Tigyi G, McKerracher L (1999) Inactivation of Rho signaling pathway promotes CNS axon regeneration. J Neurosci 19:7537–7547

Li M, Shibata A, Li C, Braun PE, McKerracher L, Roder J, Kater SB, David S (1996) Myelin-associated glycoprotein inhibits neurite/axon growth and causes growth cone collapse. J Neurosci Res 46:404–414

Liu Y, Wada R, Kawai H, Sango K, Deng C, Tai T, McDonald MP, Araujo K, Crawley JN, Bierfreund U, Sandhoff K, Suzuki K, Proia RL (1999) A genetic model of substrate

deprivation therapy for a glycosphingolipid storage disorder. J Clin Invest 103:497–505
Luo L (2000) Rho GTPases in neuronal morphogenesis. Nat Rev Neurosci 1:173–180
McKeon RJ, Schreiber RC, Rudge JS, Silver J (1991) Reduction of neurite outgrowth in a model of glial scarring following CNS injury is correlated with the expression of inhibitory molecules on reactive astrocytes. J Neurosci 11:3398–3411
McKeon RJ, Jurynec MJ, Buck CR (1999) The chondroitin sulfate proteoglycans neurocan and phosphacan are expressed by reactive astrocytes in the chronic CNS glial scar. J Neurosci 19:10778–10788
McKerracher L, David S, Jackson DL, Kottis V, Dunn RJ, Braun PE (1994) Identification of myelin-associated glycoprotein as a major myelin-derived inhibitor of neurite growth. Neuron 13:805–811
Merkler D, Metz GA, Raineteau O, Dietz V, Schwab ME, Fouad K (2001) Locomotor recovery in spinal cord-injured rats treated with an antibody neutralizing the myelin-associated neurite growth inhibitor Nogo-A. J Neurosci 21:3665–3673
Moon LD, Asher RA, Rhodes KE, Fawcett JW (2002) Relationship between sprouting axons, proteoglycans and glial cells following unilateral nigrostriatal axotomy in the adult rat. Neuroscience 109:101–117
Mukhopadhyay G, Doherty P, Walsh FS, Crocker PR, Filbin MT (1994) A novel role for myelin-associated glycoprotein as an inhibitor of axonal regeneration. Neuron 13:757–767
Niederost BP, Zimmermann DR, Schwab ME, Bandtlow CE (1999) Bovine CNS myelin contains neurite growth-inhibitory activity associated with chondroitin sulfate proteoglycans. J Neurosci 19:8979–8989
Poltorak M, Sadoul R, Keilhauer G, Landa C, Fahrig T, Schachner M (1987) Myelin-associated glycoprotein, a member of the L2/HNK-1 family of neural cell adhesion molecules, is involved in neuron-oligodendrocyte and oligodendrocyte-oligo-dendrocyte interaction. J Cell Biol 105:1893–1899
Prinjha R, Moore SE, Vinson M, Blake S, Morrow R, Christie G, Michalovich D, Simmons DL, Walsh FS (2000) Inhibitor of neurite outgrowth in humans. Nature 403:383–384
Raineteau O, Fouad K, Noth P, Thallmair M, Schwab ME (2001) Functional switch between motor tracts in the presence of the mAb IN-1 in the adult rat. Proc Natl Acad Sci USA 98:6929–6934
Schäfer M, Fruttiger M, Montag D, Schachner M, Martini R (1996) Disruption of the gene for the myelin-associated glycoprotein improves axonal regrowth along myelin in C57BL/Wld^s mice. Neuron 16:1107–1113
Schnaar RL (2000) Glycobiology of the nervous system. In: Ernst B, Hart GW, Sinaÿ P (eds) Carbohydrates in chemistry and biology, part II. Biology of saccharides. Wiley-VCH, Weinheim, pp 1013–1027
Schwab ME (2002) Repairing the injured spinal cord. Science 295:1029–1031
Schwab ME, Caroni P (1988) Oligodendrocytes and CNS myelin are nonpermissive substrates for neurite growth and fibroblast spreading in vitro. J Neurosci 8:2381–2393
Schwab ME, Thoenen H (1985) Dissociated neurons regenerate into sciatic but not optic nerve explants in culture irrespective of neurotrophic factors. J Neurosci 5:2415–2423
Sekhon LH, Fehlings MG (2001) Epidemiology, demographics, and pathophysiology of acute spinal cord injury. Spine 26:S2–S12

Selles-Navarro I, Ellezam B, Fajardo R, Latour M, McKerracher L (2001) Retinal ganglion cell and nonneuronal cell responses to a microcrush lesion of adult rat optic nerve. Exp Neurol 167:282–289

Sheikh KA, Sun J, Liu Y, Kawai H, Crawford TO, Proia RL, Griffin JW, Schnaar RL (1999) Mice lacking complex gangliosides develop Wallerian degeneration and myelination defects. Proc Natl Acad Sci USA 96:7532–7537

Snow DM, Lemmon V, Carrino DA, Caplan AI, Silver J (1990) Sulfated proteoglycans in astroglial barriers inhibit neurite outgrowth in vitro. Exp Neurol 109:111–130

Song H, Ming G, He Z, Lehmann M, Tessier-Lavigne M, Poo M (1998) Conversion of neuronal growth cone responses from repulsion to attraction by cyclic nucleotides. Science 281:1515–1518

Spillmann AA, Bandtlow CE, Lottspeich F, Keller F, Schwab ME (1998) Identification and characterization of a bovine neurite growth inhibitor (bNI-220). J Biol Chem 273:19283–19293

Stoeckli ET, Landmesser LT (1998) Axon guidance at choice points. Curr Opin Neurobiol 8:73–79

Strenge K, Brossmer R, Ihrig P, Schauer R, Kelm S (2001) Fibronectin is a binding partner for the myelin-associated glycoprotein (siglec-4a). FEBS Lett 499:262–267

Sun J, Schnaar RL (2000) Myelin-associated glycoprotein is selectively and progressively downregulated in the central and peripheral nervous systems of complex ganglioside knockout mice. Glycobiology 10:1087–1088

Svennerholm L (1994) Designation and schematic structure of gangliosides and allied glycosphingolipids. Prog Brain Res 101:xi–xiv

Tettamanti G, Bonali F, Marchesini S, Zambotti V (1973) A new procedure for the extraction, purification and fractionation of brain gangliosides. Biochim Biophys Acta 296:160–170

Trapp BD (1990) Myelin-associated glycoprotein. Location and potential functions. Ann NY Acad Sci 605:29–43

Trapp BD, Andrews SB, Cootauco C, Quarles R (1989) The myelin-associated glycoprotein is enriched in multivesicular bodies and periaxonal membranes of actively myelinating oligodendrocytes. J Cell Biol 109:2417–2426

Van Vactor D, Flanagan JG (1999) The middle and the end: slit brings guidance and branching together in axon pathway selection. Neuron 22:649–652

Vinson M, Strijbos PJ, Rowles A, Facci L, Moore SE, Simmons DL, Walsh FS (2001) Myelin-associated glycoprotein interacts with ganglioside GT1b: a mechanism for neurite outgrowth inhibition. J Biol Chem 276:20280–20285

Vyas AA, Patel HV, Fromholt SE, Heffer-Lauc M, Vyas KA, Dang J, Schachner M, Schnaar RL (2002) Gangliosides are functional nerve cell ligands for myelin-associated glycoprotein (MAG), an inhibitor of nerve regeneration. Proc Natl Acad Sci USA 99:8412–8417

Weiss MD, Luciano CA, Quarles RH (2001) Nerve conduction abnormalities in aging mice deficient for myelin-associated glycoprotein. Muscle Nerve 24:1380–1387

Yang LJS, Zeller CB, Shaper NL, Kiso M, Hasegawa A, Shapiro RE, Schnaar RL (1996) Gangliosides are neuronal ligands for myelin-associated glycoprotein. Proc Natl Acad Sci USA 93:814–818

Yin X, Crawford TO, Griffin JW, Tu P, Lee VM, Li C, Roder J, Trapp BD (1998) Myelin-associated glycoprotein is a myelin signal that modulates the caliber of myelinated axons. J Neurosci 18:1953–1962

Note added in proof: Since writing this review, three notable advances have been reported that are worthy of additional note. First, a polypeptide antagonist targeting the Nogo-66 high affinity inhibitory domain was developed and has proven effective in enhancing recovery in an in vivo spinal cord injury model in rats (GrandPré et al. 2002). Second, in addition to Nogo, both MAG (Liu et al. 2002; Domeniconi et al. 2002) and oligodendrocyte-myelin glycoprotein (Wang et al. 2002b) bind to NgR as one potential mechanism to inhibit neurite outgrowth. Finally, the p75 neurotrophin receptor mediates the inhibitory effects of Nogo, MAG and oligodendrocytemyelin glycoprotein, perhaps as a common "co-receptor" (Yamashita et al. 2002; Wang et al. 2002a).

Domeniconi M, Cao Z, Spencer T, Sivasankaran R, Wang K, Nikulina E, Kimura N, Cai H, Deng K, Gao Y, He Z, Filbin M (2002) Myelin-associated glycoprotein interacts with the nogo66 receptor to inhibit neurite outgrowth. Neuron 35:283–290

GrandPre T, Li S, Strittmatter SM (2002) Nogo-66 receptor antagonist peptide promotes axonal regeneration. Nature 417:547–551

Liu BP, Fournier A, GrandPre T, Strittmatter SM (2002) Myelin-Associated Glycoprotein as a Functional Ligand for the Nogo-66 Receptor. Science 297:1190–1193

Wang KC, Kim JA, Sivasankaran R, Segal R, He Z (2002a) P75 interacts with the Nogo receptor as a co-receptor for Nogo, MAG and OMgp. Nature 420:74–78

Wang KC, Koprivica V, Kim JA, Sivasankaran R, Guo Y, Neve RL, He Z (2002b) Oligodendrocyte-myelin glycoprotein is a Nogo receptor ligand that inhibits neurite outgrowth. Nature 417:941–944

Yamashita T, Higuchi H, Tohyama M (2002) The p75 receptor transduces the signal from myelin-associated glycoprotein to Rho. J Cell Biol 157:565–570

Subject Index

Printing (Computer to Plate): Saladruck Berlin
Binding: Stürtz AG, Würzburg